Applications of Ionic Liquids in the Oil Industry: Towards A Sustainable Industry

Authored by

Rafael Martínez Palou

&

Natalya V. Likhanova

Dirección de Investigación en Transformación de Hidrocarburos
Instituto Mexicano del Petróleo. Eje Central Lázaro Cárdenas Norte 152, 07730
Mexico City
Mexico

Applications of Ionic Liquids in the Oil Industry:

Towards A Sustainable Industry

Authors: Rafael Martínez Palou and Natalya V. Likhanova

ISBN (Online): 978-981-5079-57-9

ISBN (Print): 978-981-5079-58-6

ISBN (Paperback): 978-981-5079-59-3

Published by Bentham Science Publishers Pte. Ltd. Singapore. All Rights Reserved.

First published in 2023.

need for a court order if at any point you breach any terms of this License Agreement. In no event will any delay or failure by Bentham Science Publishers in enforcing your compliance with this License Agreement constitute a waiver of any of its rights.

3. You acknowledge that you have read this License Agreement, and agree to be bound by its terms and conditions. To the extent that any other terms and conditions presented on any website of Bentham Science Publishers conflict with, or are inconsistent with, the terms and conditions set out in this License Agreement, you acknowledge that the terms and conditions set out in this License Agreement shall prevail.

Bentham Science Publishers Pte. Ltd.
80 Robinson Road #02-00
Singapore 068898
Singapore
Email: subscriptions@benthamscience.net

CONTENTS

FOREWORD

Ionic liquids represented a real revolution in Chemical Sciences in recent decades. They have attracted enormous research efforts due to their unique properties, with the potential to replace a range of small- and large-scale processes in which technological processes present severe problems of efficiency, high toxicity, and sustainability.

The Oil Industry has not been an exception, significant advances have been made in the last two decades in new alternatives to solve technical problems with the use of ionic liquids. In fact, the alkylation process for high-quality gasoline production is one of the remarkable industrial success stories that have demonstrated that ILs (Chapter 9 of this book) have the potential to solve safety, environmental and processing issues present in oil refineries.

The Mexican Petroleum Institute (IMP), which I had the opportunity to visit a few years ago, is a research center that has devoted relevant efforts to solving technological problems in the oil industry, being the leading institution in its country in the number of patents granted and their applications. Over the years, IMP has dedicated important resources to improve the technological processes of Petróleos Mexicanos with the use of ILs as the reader will learn through this book that can be a source of inspiration and consultation for students, academics, and researchers in the area.

With very best wishes for an enjoyable and fruitful reading.

Dr. Rafael Luque

2018, 2019 and 2020 Highly Cited Researcher.

PREFACE

Ionic liquids (ILs) are ionic organic compounds, which unlike inorganic ionic compounds (salts), present, in general, very low melting points (below 100°C by and large) and in other cases, they are liquid at ambient temperature with negligible vapor pressure (non-volatile like common organic solvents), slight corrosive nature, low flammability and high chemical stability. These and other properties have positioned such compounds as "environmentally friendly" and drawn the researchers' attention, exploring a number of applications in different chemistry fields like that of the oil industry.

In the present compendium, some of the works published by the Mexican Petroleum Institute (In Spanish: Instituto Mexicano del Petróleo, IMP) are reviewed. The IMP is a public institution that has been devoted to carry out research projects aimed at providing solutions to the Mexican Oil Industry since 1965, the year when it was founded, and the synthesis and field application of ILs for dealing with the technical challenges faced by such national industry represent some of the current works developed at this research center. General aspects and recent bibliography of different topics are reviewed; in addition, IMP contributions by means of scientific papers and granted patents on ILs synthesized and used to solve technological problems found in the Mexican oil reservoirs are discussed. In this context, the removal of solid, liquid, and gaseous pollutants and the breaking of emulsions that are formed naturally between crude oil and water, which increases the oil viscosity and makes the transport of heavy and extra heavy crude oil difficult, are drawbacks that can be attacked by employing chemical compounds to control water in mature fields. ILs can be used to inhibit corrosion (corrosion inhibitors, CIs), as inhibitors of the formation of methane hydrates in deepwater wells, and as catalysts to obtain alkylate gasoline by the reaction between isobutane and butene.

This book is addressed to passionate organic chemistry researchers interested in the wide universe of ILs and more specifically to experts in research works focused on the synthesis and use of chemical compounds to support and help the Oil Industry be safer and more sustainable.

CONSENT FOR PUBLICATION

Not applicable.

CONFLICT OF INTEREST

The authors declare no conflict of interest, financial or otherwise.

ACKNOWLEDGEMENTS

The authors are grateful for the facilities granted by IMP for the writing of this chapter under project Y.62011 and thank CONACyT for financial support through project CF19-191973.

Rafael Martínez Palou

&

Natalya V. Likhanova
Dirección de Investigación en Transformación de Hidrocarburos
Instituto Mexicano del Petróleo. Eje Central Lázaro Cárdenas Norte 152, 07730
Mexico City
Mexico

CHAPTER 1

Structure, Properties and Applications of Ionic Liquids

Abstract: ILs have attracted the attention of researchers in recent decades. The number of applications in which these unusual compounds show good performance has grown dramatically in the last century. This chapter presents an overview of ionic liquids, their structure, properties and general applications that have made them one of the families of chemicals to which most research efforts have been devoted.

Keywords: Applications, Catalysis, Dissolution, Electrochemistry, Energy storage, Extraction, Ionic liquids, Properties, Polymerization reactions, Solvent, Synthesis, Separations, Synthesis of nanomaterials.

INTRODUCTION

Over the last two decades, ILs have strongly drawn the attention of the scientific community due to their interesting physical properties [1, 2] and applications as solvents with exceptional properties in organic synthesis [3 - 6], catalysis [7 - 10], biocatalysis [11 - 13], liquid-liquid separations [14], extraction [15 - 19], dissolution, [20-23] synthesis of nanomaterials [24], polymerization reactions [25, 26], electrochemistry [27, 28], and energy storage [29].

ILs are ionic compounds in which at least the cation is of organic type and have the particularity of being liquid compounds at ambient temperature or close to it (< 100°C), which makes them different from other ionic compounds or molten salts that display very high melting points (> 800°C). Fig. (**1.1**) shows the general structure of the most common cations present in ILs.

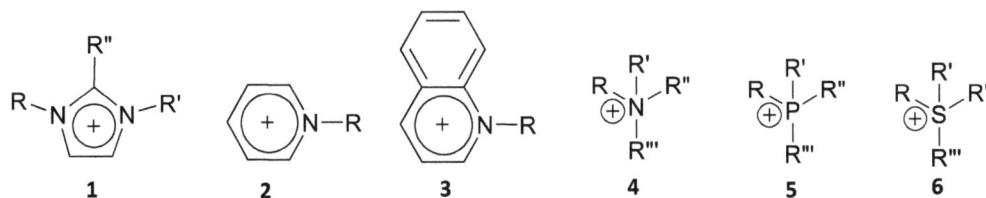

Fig. (**1.1**). Some typical IL cations. R, R', R", R''' represent alkyl, benzylic or alkyl-functionalized chains.

Rafael Martínez Palou & Natalya V. Likhanova

The cations can be of the heterocyclic type, derived from imidazole (**1**), pyridine (**2**) or quinoline (**3**) or aliphatic compounds such as quaternary compounds derived from amines (**4**), phosphorous (**5**) or sulfur (**6**) compounds. ILs have the special feature of displaying a heteroatom (nitrogen, phosphorous or sulfur) with a positive charge or electron deficiency, which in the case of aromatic derivatives, are delocalized through the ring.

In the case of anions, they can be inorganic or organic and their type affects significantly the physicochemical properties of the ILs [30].

Some of the most common ions found in ILs are as follows: Cl^-, Br^-, $[BF_4]^-$, $[PF_6]^-$, $[SbF_6]^-$, $[AlCl_4]^-$, $[FeCl_4]^-$, $[AuCl_4]^-$, $[InCl_4]^-$, $[NO_3]^-$, $[NO_2]^-$, $[SO_4]^-$, $[SCN]^-$, $[AcO]^-$, $[N(OTf)_2]^-$, $[CF_3CO_2]^-$, $[CF_3SO_3]^-$, $[PhCOO]^-$, $[C(CN)_2]^-$, $[RSO_4]^-$ and $[OTs]^-$.

The possible combinations between cations with different chain types (R) and anions allow the generation of more than 2 million of ILs with diverse physical and chemical properties [31]; some of the characteristics that make ILs so attractive in different chemical areas are the following:

Negligible vapor pressure. For this reason, ILs are considered as environmentally friendly solvents and exceptional substituents of common organic solvents, which in most cases are volatile, toxic and handled in high volumes in industrial processes.

Not flammable. This property makes them safe to be handled.

Excellent catalyst properties. The catalytic properties of these compounds are exceptional and the number of examples featuring processes where ILs have worked as catalysts is increasing exponentially in the scientific literature.

High ionic conductivity. The structure of both the cation and anion considerably We agree with the proposed change the ionic conductivity of ILs, which in general is very high.

Wide electrochemical potential window. Thanks to this feature, numberless applications in electrochemical processes are possible.

Broad thermal stability interval. For the same organic cation, the thermal stability can vary within a more or less wide interval; for this reason, these ions can be employed in processes that take place at relatively high temperatures (between 200 and 400 °C).

Variable dissolving properties. Wide range variability of the properties to dissolve organic compounds or to be dissolved in common organic solvents. The structure of both the cation and anion affects considerably the solvent properties of ILs.

Easily recyclable. ILs can be purified and reused for various cycles for many applications without altering significantly their properties or activity. The regeneration process is generally carried out by washing with conventional organic solvents and subsequent vacuum drying.

Synthesis of ILs

In general, ILs are synthesized by means of nucleophilic substitution reactions through which an alkyl halide reacts with a heteroatom in a heterocyclic or aliphatic compound, where the free electron pair from such heteroatom is involved in the formation of a new heteroatom-carbon bond, thus generating electron deficiency in the heteroatom in question.

The classical synthesis methodology of ILs occurs through the alkylation of a heteroatom with short-chain alkyl halides; for this reason, in general, at the first synthesis stage, the ILs present a halide as anion. At the second stage, the anion can be exchanged or modified through either a metathesis or acid-base reaction.

The cation in the ILs can be symmetric or asymmetric. In the case of the symmetric ILs, the reaction includes a previous stage at which the heteroatom-hydrogen bond is broken through the treatment of the heterocycles with a strong base (sodium hydride, NaH, in most cases).

The synthesis requires heating conditions under reflux with or without the presence of a solvent. The reaction time will depend mainly on the reactivity of the alkyl halides, and according to them, the reaction time can be from 24 to 72 h.

Since ILs are not volatile (practically negligible vapor pressure), purification cannot be carried out by distillation and, in general, it is performed through washings employing organic solvents capable of eliminating soluble impurities without dissolving the IL.

The preparation methods of ILs by conventional heating require many reflux hours in organic solvents, however, in the last years, synthesis methodologies using microwaves, with which both alkylation and metathesis reactions are accelerated dramatically, have been described. The microwave synthesis methodology of ILs became very popular due to the high product yields in a few reaction minutes [32]. Likewise, ILs have been a very useful auxiliary tool for microwave organic synthesis [31, 33].

Varma *et al.* described for the first time the microwave synthesis of ILs of the imidazole and bis-imidazolium type by means of the reaction between 1-methylimidazole and alkyl halides or dihalides in an open system without using a solvent. The ILs were produced in less than 2 min with yields above 70% [34]. These researchers also published an efficient methodology for the synthesis of these compounds using ultrasound as an alternative energy source [35].

Fig. (**1.2**) shows a simplified reaction diagram for the synthesis of ILs, both symmetric and asymmetric, employing conventional heating (Δ), ultrasound [)))] or microwaves (MW) as nonconventional heating sources. The schematic representation displays the synthesis from imidazole, which is one of the most used starting materials, but in general, the synthesis procedure is valid for most ILs described in the scientific literature containing the cations **2-6** described in the general structures [36 - 39].

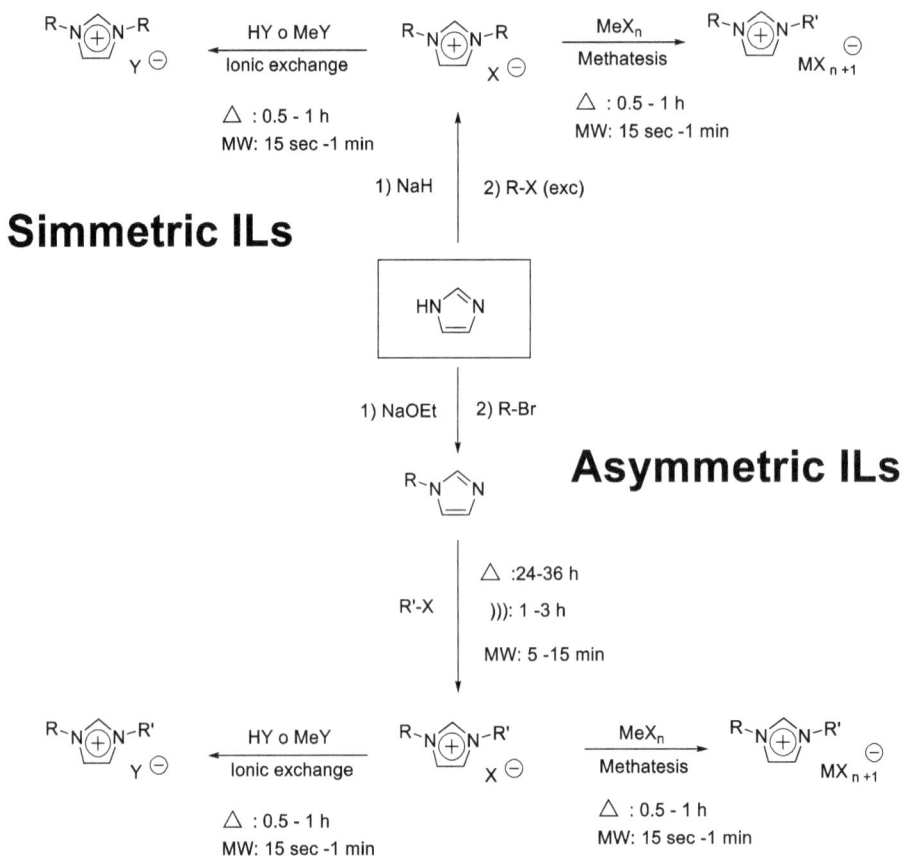

Fig. (1.2). General synthesis methodology of ILs from imidazole.

As it can be observed in Fig. (**1.2**), the synthesis time of the ILs is reduced considerably by employing a microwave piece of equipment. According to the aforementioned, it would be recommendable that such a piece be at hand for the synthesis of the ILs that are intended to be evaluated.

Once the ILs are synthesized and purified, they are submitted to a structural characterization process in order to obtain unequivocal information of the synthesized compound and its purity by means of techniques such as nuclear magnetic resonance (NMR), infrared spectroscopy and mass spectrometry.

Technology Development Using ILs

Since the 1980s, the interest in the use of ILs as solvents in chemical processes has increased notably and since then, a large number of applications have emerged. Many of these research works have focused on the employment of ILs to create biphasic systems for alkylation and acetylation reactions [40].

The ILs can be classified as "design solvents", for by varying the characteristics of the implied ions, millions of different combinations can be synthesized; which is an immense amount in comparison with the less than 300 most used organic solvents in the chemical industry. Due to the excellent hybrid properties of ILs, which stem from their organic-inorganic nature (thinking of a cleaner chemical industry), the most varied uses of ILs in different application areas have been suggested and studied, as described in Table **1.1**.

Table 1.1. Some applications of the ILs.

Application Area	Some Possible Applications
Energy	• Fuel cells • Photovoltaic cells • Light emission electrochemical cells • Electrolytes for lithium batteries • Electrolytes for solar cells
Chemistry	• Organic synthesis • Chiral synthesis • Polymerization • Catalysis • Electrosynthesis of conducting polymers
Biotechnology	• Biocatalysis • Purification of proteins

(Table 1.1) cont.....

Application Area	Some Possible Applications
Chemical Engineering	• Extraction with supercritical fluids • Separation processes • Membranes • Extractive distillation • Cleaner fuels
Other	• Nanoparticles • Liquid crystals • Additives

Large-scale Applications in the Oil Industry

As for the applications of ILs in the Oil Industry, some recent developments have been described; for example, the French Petroleum Institute patented the use of ILs as solvents for alkylation, polymerization, and catalysts for the Diels-Alder reaction [41].

The company BP Chemicals tested the use of pyridinium or imidazole chloride in combination with an alkyl aluminum halide (R_nAlX_{3-n}) as IL for the polymerization of butane [42, 43]. On the other hand, Akzo Nobel described a process for the formation of alkylbenzenes employing ILs [44, 45]. Exxon patented a process using ILs for extracting aromatic compounds from a hydrocarbon mixture, where triethylammonium dihydroxybenzoate salts were employed [46].

One of the applications that have had the largest scaling of a productive process employing ILs is the technology for gasoline production by alkylation through the reaction between isobutane and butenes. This technology has been widely studied to replace the conventional acid catalysts (H_2SO_4 and HF) by ILs [47, 48]. These ILs contain transition metals in their anion and have been used as catalysts of the alkylation reaction between light olefins and hydrocarbons, typically between 2-butene and isobutane to obtain trimethylpentane. The company PetroChina developed an alkylation process based on ILs known as Ionikylation. Currently, the company has an alkylation unit based on this technology that produces 150,000 ton/year [49 - 51]. As for Honeywell UOP, it licensed the technology known as Isoalky™, which was developed by Chevron, at the test stage in Salt Lake City based on a chloroaluminate-type catalyst [52].

Another advantage associated with the use of ILs as solvents in chemical reactions is that they require milder temperature conditions than when conventional solvents are employed, which implies as a consequence, the reduction of energy and environmental costs [53 - 55]. For example, the reactions, where ILs are used as catalysts, occur at lower temperatures and with higher yields. The Friedel-Crafts reaction, which plays a key role in the oil cracking process, is carried out with conventional solvents at 80 ° C, takes 8 h and has a yield of 80%; in contrast, the same reaction using ILs is performed at 0 °C, occurs in 30 seconds with a yield of 98% and a product that is purer and homogeneous [56 - 60].

At present, ILs have widened their applications in processes at a large scale [61] and many of them are commercially available [62]. Recently, some current and future applications have been reviewed [63, 64].

CONCLUDING REMARKS

As we have seen in the present chapters, ILs have such interesting, varied, and unique properties, besides being such a wide and diverse family of compounds that make them a focus of attention for their application in different scientific fields, such as organic synthesis, catalysis, biocatalysts, separations, extraction, dissolution, synthesis of nanomaterials, polymerization reactions, electrochemistry, and energy storage. As you can see by reading this book, the application of ionic liquids has been very extensive in a wide variety of applications in the Oil Industry, from facilitating the processes of primary and improved extraction of crude oil from the deep sea to the addition of these products in the crude transport and refining processes. Some of these applications are still in the early stages of research and certain challenges remain to be resolved, while other results have already been tested on an industrial scale.

REFERENCES

[1]   Pernak, J. Ionic liquids. Compounds for the 21ˢᵗ Century. *Przem. Chem.,* **2003,** *82,* 521-524.

[2]   Forsyth, S.A.; Pringle, J.M.; MacFarlane, D.R. Ionic liquids-an overview. *Aust. J. Chem.,* **2004,** *57*(2), 113-119.
[http://dx.doi.org/10.1071/CH03231]

[3]   Wasserscheid, P.; Keim, W., Eds. Ionic Liquids in Synthesis. Wiley-VCH: Wenheim, **2004.**

[4]   Rogers, R.D.; Seddon, K.R., Eds. *Ionic Liquids as Green Solvent: Progress and Prospects*; ACS: Boston, **2003.**
[http://dx.doi.org/10.1021/bk-2003-0856]

[5]   Hallett, J.P.; Welton, T. Room-temperature ionic liquids: solvents for synthesis and catalysis. 2. *Chem. Rev.,* **2011,** *111*(5), 3508-3576.
[http://dx.doi.org/10.1021/cr1003248] [PMID: 21469639]

[6]   Martins, M.A.P.; Frizzo, C.P.; Moreira, D.N.; Zanatta, N.; Bonacorso, H.G. Ionic liquids in heterocyclic synthesis. *Chem. Rev.,* **2008,** *108*(6), 2015-2050.
[http://dx.doi.org/10.1021/cr078399y] [PMID: 18543878]

[7]   Zhao, D.; Wu, M.; Kou, Y.; Min, E. Ionic liquids: applications in catalysis. *Catal. Today,* **2002,** *74*(1-2), 157-189.
[http://dx.doi.org/10.1016/S0920-5861(01)00541-7]

[8]   Welton, T. Room-Temperature Ionic Liquids. Solvents for Synthesis and Catalysis. *Chem. Rev.,* **1999,**

99(8), 2071-2084.
[http://dx.doi.org/10.1021/cr980032t] [PMID: 11849019]

[9]   Pârvulescu, V.I.; Hardacre, C. Catalysis in ionic liquids. *Chem. Rev.,* **2007**, *107*(6), 2615-2665.
       [http://dx.doi.org/10.1021/cr050948h] [PMID: 17518502]

[10]   Qiao, Y.; Ma, W.; Theyssen, N.; Chen, C.; Hou, Z. Temperature-Responsive Ionic Liquids:
       Fundamental Behaviors and Catalytic Applications. *Chem. Rev.,* **2017**, *117*(10), 6881-6928.
       [http://dx.doi.org/10.1021/acs.chemrev.6b00652] [PMID: 28358505]

[11]   Cull, S.G.; Holbrey, J.D.; Vargas-Mora, V.; Seddon, K.R.; Lye, G.J. Room-temperature ionic liquids
       as replacements for organic solvents in multiphase bioprocess operations. *Biotechnol. Bioeng.,* **2000**,
       69(2), 227-233.
       [http://dx.doi.org/10.1002/(SICI)1097-0290(20000720)69:2<227::AID-BIT12>3.0.CO;2-0]   [PMID:
       10861402]

[12]   Sheldon, R.A.; Lau, R.M.; Sorgedrager, M.J.; van Rantwijk, F.; Seddon, K.R. Biocatalysis in ionic
       liquids. *Green Chem.,* **2002**, *4*(2), 147-151.
       [http://dx.doi.org/10.1039/b110008b]

[13]   Itoh, T. Ionic Liquids as Tool to Improve Enzymatic Organic Synthesis. *Chem. Rev.,* **2017**, *117*(15),
       10567-10607.
       [http://dx.doi.org/10.1021/acs.chemrev.7b00158] [PMID: 28745876]

[14]   Ventura, S.P.M.; e Silva, F.A.; Quental, M.V.; Mondal, D.; Freire, M.G.; Coutinho, J.A.P. Ionic-
       Liquid-Mediated Extraction and Separation Processes for Bioactive Compounds: Past, Present, and
       Future Trends. *Chem. Rev.,* **2017**, *117*(10), 6984-7052.
       [http://dx.doi.org/10.1021/acs.chemrev.6b00550] [PMID: 28151648]

[15]   Blanchard, L.A.; Hancu, D.; Beckman, E.J.; Brennecke, J.F. Green processing using ionic liquids and
       CO2. *Nature,* **1999**, *399*(6731), 28-29.
       [http://dx.doi.org/10.1038/19887]

[16]   Bösmann, A.; Datsevich, L.; Jess, A.; Lauter, A.; Schmitz, C.; Wasserscheid, P. Deep desulfurization
       of diesel fuel by extraction with ionic liquids. *Chem. Commun. (Camb.),* **2001**, (23), 2494-2495.
       [http://dx.doi.org/10.1039/b108411a] [PMID: 12240031]

[17]   Zhang, S.; Conrad Zhang, Z. Novel properties of ionic liquids in selective sulfur removal from fuels at
       room temperature. *Green Chem.,* **2002**, *4*(4), 376-379.
       [http://dx.doi.org/10.1039/b205170m]

[18]   Zhang, S.; Zhang, Q.; Zhang, Z.C. Extractive Desulfurization and Denitrogenation of Fuels Using
       Ionic Liquids. *Ind. Eng. Chem. Res.,* **2004**, *43*(2), 614-622.
       [http://dx.doi.org/10.1021/ie030561+]

[19]   Sun, X.; Luo, H.; Dai, S. Ionic liquids-based extraction: a promising strategy for the advanced nuclear
       fuel cycle. *Chem. Rev.,* **2012**, *112*(4), 2100-2128.
       [http://dx.doi.org/10.1021/cr200193x] [PMID: 22136437]

[20]   Swatloski, R.P.; Spear, S.K.; Holbrey, J.D.; Rogers, R.D. Dissolution of Cellose with Ionic Liquids. *J.
       Am. Chem. Soc.,* **2002**, *124*(18), 4974-4975.
       [http://dx.doi.org/10.1021/ja025790m] [PMID: 11982358]

[21]   Merino, O.; Fundora, G.; Luque, R.; Martínez-Palou, R. Understanding microwave-assisted lignin
       solubilization in Protic Ionic Liquids with multiaromatic imidazolium cation. Study of lignin
       solubilization. *ACS Sustain. Chem.& Eng.,* **2018**, *6*, 4122-4129.
       [http://dx.doi.org/10.1021/acssuschemeng.7b04535]

[22]   Merino, O.; Cerón-Camacho, R.; Martínez-Palou, R. Microwave-assisted lignin solubilization.
       Synthesis on Protic Ionic Liquids from multiaromatic imidazole with inorganic anion. *Waste Biomass
       Valorization,* **2020**, *11*, 6585-6593.
       [http://dx.doi.org/10.1007/s12649-016-9612-3]

[23] Verma, C., Mishra, A., Chauhan, S., Verma, P., Srivastava, V., Quraishi, M. A., & Ebenso, E. E. Dissolution of cellulose in ionic liquids and their mixed cosolvents: A review. *Sustain. Chem. Pharm.,* **2019**, *13*, 100162.

[24] Cooper, E.R.; Andrews, C.D.; Wheatley, P.S.; Webb, P.B.; Wormald, P.; Morris, R.E. Ionic liquids and eutectic mixtures as solvent and template in synthesis of zeolite analogues. *Nature,* **2004**, *430*(7003), 1012-1016.
[http://dx.doi.org/10.1038/nature02860] [PMID: 15329717]

[25] Kubisa, P. Application of ionic liquids as solvents for polymerization processes. *Prog. Polym. Sci.,* **2004**, *29*(1), 3-12.
[http://dx.doi.org/10.1016/j.progpolymsci.2003.10.002]

[26] Vygodskii, Y.S.; Lozinskaya, E.I.; Shaplov, A.S.; Lyssenko, K.A.; Antipin, M.Y.; Urman, Y.G. Implementation of ionic liquids as activating media for polycondensation processes. *Polymer (Guildf.),* **2004**, *45*(15), 5031-5045.
[http://dx.doi.org/10.1016/j.polymer.2004.05.025]

[27] Yang, C.; Sun, Q.; Qiao, J.; Li, Y. Ionic Liquid Doped Polymer Light-Emitting Electrochemical Cells. *J. Phys. Chem. B,* **2003**, *107*(47), 12981-12988.
[http://dx.doi.org/10.1021/jp034818t]

[28] Quinn, B.M.; Ding, Z.; Moulton, R.; Bard, A.J. Novel Electrochemical Studies of Ionic Liquids. *Langmuir,* **2002**, *18*(5), 1734-1742.
[http://dx.doi.org/10.1021/la011458x]

[29] Watanabe, M.; Thomas, M.L.; Zhang, S.; Ueno, K.; Yasuda, T.; Dokko, K. Application of Ionic Liquids to Energy Storage and Conversion Materials and Devices. *Chem. Rev.,* **2017**, *117*(10), 7190-7239.
[http://dx.doi.org/10.1021/acs.chemrev.6b00504] [PMID: 28084733]

[30] Hettige, J.J.; Kashyap, H.K.; Annapureddy, H.V.R.; Margulis, C.J. Anions, the Reporters of Structure in Ionic Liquids. *J. Phys. Chem. Lett.,* **2013**, *4*(1), 105-110.
[http://dx.doi.org/10.1021/jz301866f] [PMID: 26291220]

[31] Martínez-Palou, R. Microwave-assisted synthesis using ionic liquids. *Mol. Divers.,* **2010**, *14*(1), 3-25.
[http://dx.doi.org/10.1007/s11030-009-9159-3] [PMID: 19507045]

[32] Deetlefs, M.; Seddon, K.R. Improved preparations of ionic liquids using microwave irradiationThis work was presented at the Green Solvents for Catalysis Meeting held in Bruchsal, Germany, 13–16th October 2002. *Green Chem.,* **2003**, *5*(2), 181-186.
[http://dx.doi.org/10.1039/b300071k]

[33] Mallakpour, S.; Rafiee, Z. New developments in polymer science and technology using combination of ionic liquids and microwave irradiation. *Prog. Polym. Sci.,* **2011**, *36*(12), 1754-1765.
[http://dx.doi.org/10.1016/j.progpolymsci.2011.03.001]

[34] Varma, R.S.; Namboodiri, V.V. An expeditious solvent-free route to ionic liquids using microwaves. *Chem. Commun. (Camb.),* **2001**, (7), 643-644.
[http://dx.doi.org/10.1039/b101375k]

[35] Zang, H.; Su, Q.; Mo, Y.; Cheng, B.W.; Jun, S. Ionic liquid [EMIM]OAc under ultrasonic irradiation towards the first synthesis of trisubstituted imidazoles. *Ultrason. Sonochem.,* **2010**, *17*(5), 749-751.
[http://dx.doi.org/10.1016/j.ultsonch.2010.01.015] [PMID: 20194046]

[36] Ameta, G.; Pathak, A.K.; Ameta, C.; Ameta, R.; Punjabi, P.B. Sonochemical synthesis and characterization of imidazolium based ionic liquids: A green pathway. *J. Mol. Liq.,* **2015**, *211*, 934-937.
[http://dx.doi.org/10.1016/j.molliq.2015.08.009]

[37] Estager, J.; Lévêque, J.M.; Cravotto, G.; Boffa, L.; Bonrath, W.; Draye, M. One-pot and Solventless Synthesis of Ionic Liquids under Ultrasonic Irradiation. *Synlett,* **2007**, *13*, 2065-2068.

[http://dx.doi.org/10.1055/s-2007-984881]

[38]   Rogers, R.D.; Seddon, K.R., Eds. *Ionic Liquids: Industrial Applications for Green Chemistry*; ACS: Boston, **2002**.
[http://dx.doi.org/10.1055/s-2007-984881]

[39]   Hoffmann, J.; Nüchter, M.; Ondruschka, B.; Wasserscheid, P. Ionic liquids and their heating behaviour during microwave irradiation – a state of the art report and challenge to assessment. *Green Chem.,* **2003**, *5*(3), 296-299.
[http://dx.doi.org/10.1039/B212533A]

[40]   Sheldon, R.A.; van Rantwijk, F.; Machas Madeira Lau, R. Biotransformations in ionic liquids: an overview. In: *Progress and prospects*; , **2003**.
[http://dx.doi.org/10.1021/bk-2003-0856.ch016]

[41]   Olivier, H.; Hirschauer, A. Process for dienoic condensation known as the diels-alder reaction. U.S. Pat. 5. *892*, 124.

[42]   Chauvin, Y.; Einloft, S.; Olivier, H. Nickel-containing composition for catalysis and olefin dimerisation and oligomerisation process. *U.S Pat. 5, 550*, 304.

[43]   Ambler, P.W.; Hodgson, P.K.G.; Stewart, N.J. Preparation of butene polymers using an ionic liquid. U.S. Pat 5. *304*, 615.

[44]   Sherif, F.G. Linear alxylbenzene formation using low temperature ionic liquid. U.S. Pat. 5. *824*, 832.

[45]   Sherif, F. G.; Shyu, L.-J.; Greco, C. C.; Talma, A. G.; Lacroix, C. P. M. Linear alkylbenzene formation using low temperature ionic liquid and long chain alkylating agent. EP0963366A1.

[46]   Boate, D. R.; Zaworotko, M. J. Aromatic hydrocarbons separation from aromatic/non-aromatic mixtures. Eur. Pat. 0562815A1.

[47]   Timken, H.K.; Luo, H.; Chang, B.K.; Carter, E.; Cole, M. ISOALKY™ Technology: Next-Generation Alkylate Gasoline Manufacturing Process Technology Using Ionic Liquid Catalyst. In: *Commercial Applications of Ionic Liquids. Green Chemistry and Sustainable Technology*; Shiflett, M., Ed.; Springer, **2020**.
[http://dx.doi.org/10.1007/978-3-030-35245-5_2]

[48]   McCoy, M. Chevron embraces ionic liquids. *Chem. Eng. News,* **2016**, *94*, 16.

[49]   Liu, Z.; Zhang, R.; Xu, C.; Xia, R. Ionic liquid alkylation process produces high-quality gasoline. *Oil Gas J.,* **2006**, *104*, 52-56.

[50]   Petro China's first ionic liquid alkylation unit. Hydrocarbon Processing. *PetroChina's first ionic liquid alkylation unit,* **2018**. www.hydrocarbonprocessing.com

[51]   Díaz Velázquez, H.; Likhanova, N.; Aljammal, N.; Verpoort, F.; Martínez-Palou, R. New Insights into the Progress on the Isobutane/Butene Alkylation Reaction and Related Processes for High-Quality Fuel Production. A Critical Review. *Energy Fuels,* **2020**, *34*(12), 15525-15556.
[http://dx.doi.org/10.1021/acs.energyfuels.0c02962]

[52]   Affairs, C.P. Government and Public. Chevron and Honeywell Announce Start-up of World's First Commercial ISOALKYTM Ionic Liquids Alkylation Unit. https://www.chevron.com/stories/chevron-and-honeywell-announce-start-up-of-isoalky-ionic-liquids-alkylation-unit

[53]   Liu, C.Z.; Wang, F.; Stiles, A.R.; Guo, C. Ionic liquids for biofuel production: Opportunities and challenges. *Appl. Energy,* **2012**, *92*, 406-414.
[http://dx.doi.org/10.1016/j.apenergy.2011.11.031]

[54]   Ding, G.; Liu, Y.; Wang, B.; Punyapitak, D.; Guo, M.; Duan, Y.; Li, J.; Cao, Y. Preparation and characterization of fomesafen ionic liquids for reducing the risk to the aquatic environment. *New J. Chem.,* **2014**, *38*(11), 5590-5596.
[http://dx.doi.org/10.1039/C4NJ01186D]

[55]   Guo, F.; Zhang, S.; Wang, J.; Teng, B.; Zhang, T.; Fan, M. Synthesis and Applications of Ionic Liquids in Clean Energy and Environment: A Review. *Curr. Org. Chem.,* **2015**, *19*(5), 455-468.
[http://dx.doi.org/10.2174/1385272819666150114235649]

[56]   Valkenberg, M.H.; deCastro, C.; Hölderich, W.F. Friedel-Crafts acylation of aromatics catalysed by supported ionic liquids. *Appl. Catal. A Gen.,* **2001**, *215*(1-2), 185-190.
[http://dx.doi.org/10.1016/S0926-860X(01)00531-2]

[57]   Xiao, Y.; Malhotra, S.V. Friedel-Crafts acylation reactions in pyridinium based ionic liquids. *J. Organomet. Chem.,* **2005**, *690*(15), 3609-3613.
[http://dx.doi.org/10.1016/j.jorganchem.2005.04.047]

[58]   Baleizão, C.; Pires, N.; Gigante, B.; Marcelo Curto, M.J. Friedel–Crafts reactions in ionic liquids: the counter-ion effect on the dealkylation and acylation of methyl dehydroabietate. *Tetrahedron Lett.,* **2004**, *45*(22), 4375-4377.
[http://dx.doi.org/10.1016/j.tetlet.2004.03.185]

[59]   Earle, M.J.; Seddon, K.R.; Adams, C.J.; Roberts, G. Friedel–Crafts reactions in room temperature ionic liquids. *Chem. Commun. (Camb.),* **1998**, (19), 2097-2098.
[http://dx.doi.org/10.1039/a805599h]

[60]   Wang, Z.W.; Wang, L.S. Preparation of dichlorophenylphosphine *via* Friedel–Crafts reaction in ionic liquids. *Green Chem.,* **2003**, *5*(6), 737-739.
[http://dx.doi.org/10.1039/B307418H]

[61]   Chen, L.; Sharifzadeh, M.; Mac Dowell, N.; Welton, T.; Shah, N.; Hallett, J.P. Inexpensive ionic liquids: [HSO$_4$]$^-$-based solvent production at bulk scale. *Green Chem.,* **2014**, *16*(6), 3098-3106.
[http://dx.doi.org/10.1039/C4GC00016A]

[62]   Commercial Applications of Ionic Liquids. *Green Chemistry and Sustainable Technology*; Shiflett, M., Ed. Springer: Cham, Switzerland, **2020**.

[63]   Nasirpour, N.; Mohammadpourfard, M.; Zeinali Heris, S. Ionic liquids: Promising compounds for sustainable chemical processes and applications. *Chem. Eng. Res. Des.,* **2020**, *160*, 264-300. [first_page settings].
[http://dx.doi.org/10.1016/j.cherd.2020.06.006]

[64]   Greer, A.J.; Jacquemin, J.; Hardacre, C. Industrial Applications of Ionic Liquids. *Molecules,* **2020**, *25*(21), 5207-5238.
[http://dx.doi.org/10.3390/molecules25215207] [PMID: 33182328]

CHAPTER 2

Application of ILs in the Removal of Pollutants Present in Gasoline and Diesel

Abstract: This chapter presents an overview of ionic liquids application for the removal of some pollutants such as sulfur, nitrogen and others that are present in considerable concentrations in fuels such as gasoline and diesel and which must be removed because they cause major environmental problems, and which can be extracted by different liquid-liquid extraction procedures using ILs.

Keywords: Ionic liquids, Pollutants, Gasoline, Diesel, Liquid-liquid extraction, Sulfur, Nitrogen, Fluoride.

INTRODUCTION

Oil consists mainly of hydrocarbons that present high combustion efficiency; however, it is inevitably accompanied by other organic and inorganic compounds such as water, sulfur, nitrogen, oxygenated and halogenated organic compounds, resins, inorganic salts and carbon dioxide, among others. Most of these compounds can be highly pollutant to the environment and additionally, some of them diminish the combustible properties of hydrocarbons. Through the refining process of crude oil, it is possible to separate partially some contaminants and in turn, new pollutants such as those known as greenhouse effect gases, which are the main promoters of acid rain, are produced; for this reason, these compounds should be separated as exhaustively as possible from the oil derivatives generated during the refining process [1].

The sulfur content varies according to the origin of crude oil and since that many sulfur compounds vaporize within the same boiling interval of the primary product, these compounds are present too, polluting them (Table **2.1**) [2, 3].

At the industrial level, the removal of sulfur compounds is carried out by a hydrotreatment process called Hydrodesulfurization (HDS) [4, 5]. Around 40% of the total gasoline mixture comes from either atmospheric residues or vacuum distillates that produce FCC gasoline, which contributes to 85-95% of the sulfur content and olefins in the FCC effluents (Fig. **2.1**) [6 - 10].

Rafael Martínez Palou & Natalya V. Likhanova

Table 2.1. Some of the pollutants present in the different oil fractions.

Fraction	Main Pollutants
Gasoline: Naphtha, Naphtha for Fluid Catalytic Cracking (FCC).	Mercaptans (R-S-H), Sulfides and Disulfides (R-S-S-R).
Jet Fuel: Heavy naphtha, middle distillate.	Benzothiophene and its alkyl derivatives.
Diesel: middle distillate, light cycle oil (LCO).	Alkyl benzothiophenes, dibenzothiophenes and its alkyl derivatives.

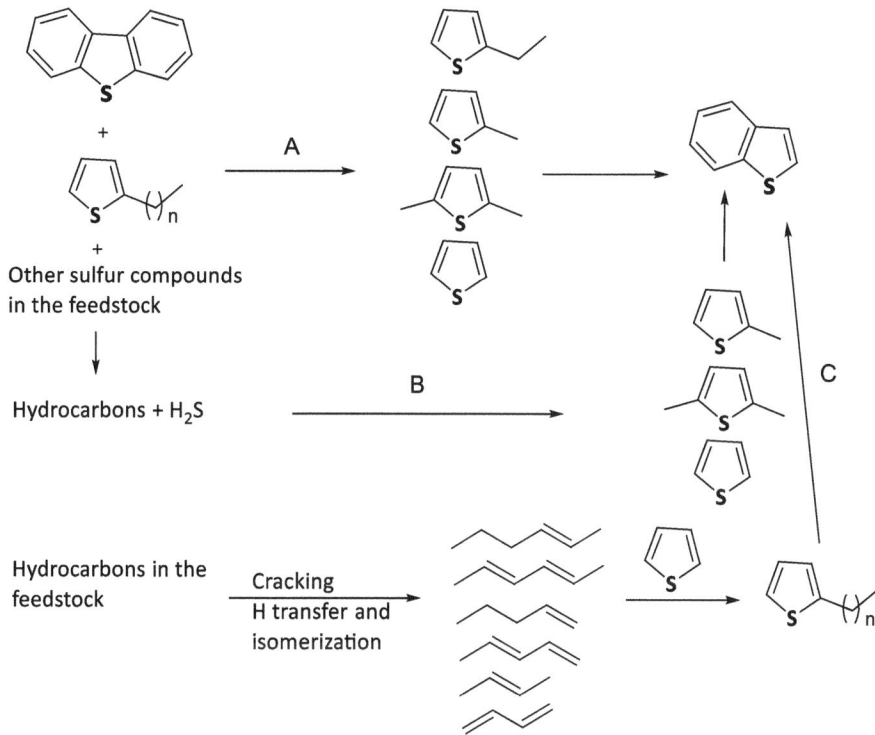

Fig. (2.1). Schematic representation of the formation and recombination of sulfur compounds through the FCC process: (A) transformation of heavy sulfur compounds in the feedstock, (B) reaction between H₂S (produced by the desulfurization of feedstock impurities) and olefins or diolefins resulting from the catalytic cracking of the feedstock, and (C) cyclization of alkylthiophenes formed during the process.

The FCC of gasoline promotes the direct combination and transformation of sulfur compounds present in the feedstock, producing many impurities [11].

Not all sulfur compounds can be eliminated through conventional techniques, and for this reason, they have to be submitted to more severe treatments. Thiophenic compounds, especially the 4,6-dialkyl-substituted compounds are difficult to be transformed into H_2S due to their chemical stability and steric hindrance of the interaction between the sulfur atom in their structure and the catalyst surface [12].

The oil refining industry must adapt itself to the environmental legislation and engine design changes, which in turn are adapted to environmental requisites. The necessity of protecting both automobile parts and industrial pieces of equipment from corrosion, the commercial opening among different countries, and release of the international oil prices are factors that have increased the demand for more and better fuels from the oil refining industry [13, 14].

To reach these goals, using the current HDS technology, higher temperatures, pressure, more efficient reactors, and more active catalysts are needed, which implies an important increase in the process cost [15].

Environmental Problems Due to the Presence of Sulfur Compounds in Fuels

The main source of atmospheric pollution is the use of fossil fuels as energy suppliers. Huge amounts of oil, gas and coal are, in the order of millions of tons, consumed every day, and the combustion residues are expelled into the atmosphere as solid particles, smoke and gases that trigger problems such as acid rain. Some studies have stated that automotive vehicles contribute to more than 90% of emissions, and for this reason, many environmental strategies are aimed at this sector.

The main pollutants associated with combustion are particles, SO_x, NO_x, CO_2, CO and hydrocarbons; in general, the industry is responsible for 55% of the sulfur dioxide (SO_2) emissions and the rest of the contribution is due fundamentally to transport [16].

Combustion gases from oil derived fuels play a major role in both acid rain, planet heating (global warming) and in the increase of the tropospheric ozone and carbon monoxide levels, which are highly toxic for human beings [17].

The main components of acid rain are formed from primary pollutants such as sulfur dioxide and nitrogen oxides through the reactions presented in Fig. (**2.2**).

The primary pollutants emitted by combustion (reaction (1), (2) and (3) suffer additional oxidation and the products can react easily with atmospheric humidity (4) and (5) and remain dissociated as part of the fog, snow, or rain, thus producing acid rain or fog.

$$SO_2 + 1/_2O_2 \longrightarrow SO_3 \qquad (1)$$
$$NO + 1/_2O_2 \longrightarrow NO_2 \qquad (2)$$
$$NO + O_3 \longrightarrow NO_3 + O_2 \qquad (3)$$
$$SO_2 + H_2O \longrightarrow H_2SO_4 \qquad (4)$$
$$3NO_2 + O_3 \longrightarrow NO_3 + O_2 \qquad (5)$$

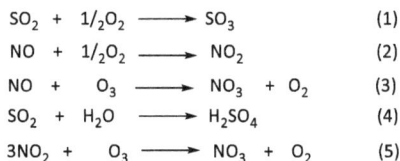

Fig. (2.2). Main SO_x and NO_x reactions that produce the acid rain.

These compounds, as part of rain and fog drops, have a short lifespan and react quickly with organic and inorganic compounds; by reacting, they are consumed, but leave behind damage that is manifested in the form of mucous irritation in humans and animals or worn out of the leaf cuticle in vegetables, favoring the entrance of pathogen microorganisms, which in turn reduces the agricultural production and damages facilities and pieces of equipment left in the open air; for these reasons, it is important that acid rain and fog be fought back by avoiding the implied emissions.

Currently, there is very strict environmental normativity in almost all oil-producing countries to limit the maximal content of sulfur, nitrogen, and other pollutants in the production of fuels from hydrocarbons [18].

The use of ILs to develop clean technologies for the Oil Industry was reviewed by an IMP work group in 2011 and described in Chapter 24 of the book: Perspectives of Ionic Liquids for Clean Oilfield Technologies. R. Martínez-Palou, P. Flores. Ionic Liquids. Theory, Properties [19].

And more recently in a revision published in 2014: Applications of Ionic liquids for Removing Pollutants from Refinery Feedstocks: A review, by R. Luque, R. Martínez-Palou. Environm. Energy Sci. 2014, 7(8), 2414-2447,20 where the state of the art of the removal of contaminants employing ILs in the Oil Industry was discussed. In this review, stood out that, to date, the environmental regulations are more restrictive every time and more frequently implemented in the oil refineries to reduce both the content of toxic fuel compounds and the negative effects related to health and the environment. However, the exhaustive removal of pollutants from refinery raw materials (for example, gasoline and diesel) is a highly challenging task that requires more ecological and sustainable technologies capable of being as efficient as the conventional methods [20].

Refineries also face big challenges to comply with the specification of fuel pollutants to achieve and produce higher fuel qualities and added-value compounds. One of the high-impact topics in this industry is the development of more ecological and sustainable processes for the removal of contaminants [21].

Recently, new no-hydrodesulfurization technologies, including the use of ILs as extraction means, have been studied in order to eliminate the pollutants from the different refining industry products. An alternative is the one called extractive desulfurization, which seems to be very attractive for this purpose due to its low energy costs, no hydrogen use, preservation of the fuel chemical structures and no need for special pieces of equipment [22].

At first, extractive desulfurization employed conventional solvents as extraction agents, where dimethyl sulfoxide, pyrimidone, imidazolidinone and polyalkylene glycol are some examples, but in the last decades, it has been shown that ILs can be a more effective tool for this purpose as described in the following section.

Application of ILs in the Desulfurization of Oil Derived Fuels

In the last years, the deep desulfurization of oil-derived fuels has strongly drawn attention due to the SOx emissions generated by the combustion of hydrocarbons. The compounds that contain sulfur also reduce the combustion efficiency of diesel fuels and poison the catalysts employed in the HDS process. For this reason, the Environmental Protection Agency (EPA) has established stricter regulations for oil refineries in order to limit the maximal sulfur content in fuels. These actions are also aimed at reducing the impact of SOx emissions during fuel combustion on human health and the environment.

The removal of organosulfur and organonitrogen compounds from fuel oils, which is carried out at the industrial level by means of the HDS and hydrodenitrogenation (HDN) processes at high temperatures (at around 350°C and above), classically requires catalysts based on metal mixtures such as cobalt/molybdenum or nickel/molybdenum. Through these processes, the rupture of C-S and C-N bonds occurs to produce mainly hydrogen sulfide and ammonia, respectively [23 - 28].

The deep desulfurization of diesel fuels is particularly challenging due to difficulties associated with the reduction of aromatic sulfur compounds whose structures and relatively complex hydrodesulfurization are shown in Fig. (2.3).

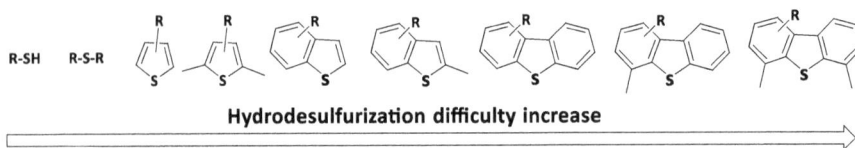

Hydrodesulfurization difficulty increase

Fig. (2.3). Some typical sulfur compounds present in hydrocarbons and their oxidation ease through the HDS process (R = short alkyl chain).

In general, the HDS process is just very effective to eliminate aliphatic- and alicyclic-type organosulfur compounds. The conversion of aromatic sulfur molecules, which include thiophenes (TS), dibenzothiophenes (DBT) and alkyl derivatives, into H$_2$S by means of HDS catalysts is a complex process and the exhaustive desulfurization of these compounds requires catalysts that must be more and more efficient and higher energy and hydrogen levels [29].

Recently, new processes and technologies that do not imply the HDS process to produce clean fuels with very low sulfur content have been studied [30 - 35]. In this context, a very promising alternative that has been referred to as extractive desulfurization (EDS) could be particularly suitable for achieving this goal.

On the other hand, liquid-liquid extraction is an industrial methodology that is widely employed to separate mixtures. The advantages of this method are the simplified operation option, mild process conditions and low energy demand. Nonetheless, the extraction efficiency depends mainly on the careful selection of the solvents for specific separation processes. Conventional solvents for extraction purposes are highly volatile, inflammable, and often dangerous and pollutants; for this reason, ILs are very attractive, in addition to their ecological potential for carrying out the liquid-liquid extraction process efficiently. Fig. (**2.4**) shows the schematic representation of the removal process of sulfur compounds present in naphtha through their migration to the IL phase that is immiscible in the hydrocarbon (HC) phase, which are part of naphtha, employing *n*-pentane as an example of one of these HCs and ethanethiol as one of the sulfur compounds that are commonly present in naphtha [36 - 42].

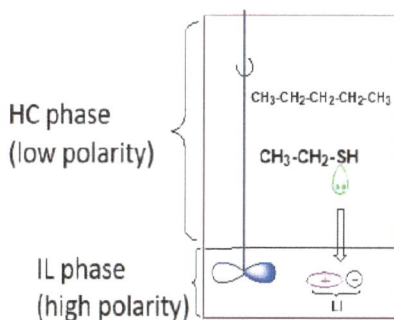

Fig. (2.4). Schematic representation of the liquid-liquid desulfurization process employing ILs.

A pioneering work on the extraction of sulfur compounds through liquid-liquid extraction employing ILs was published by Bösmann and colleagues in 2001 [43]. In this study, two ILs with aluminates as anions (imidazole methyl butyl trichloroaluminate, BuMeIMCl/AlCl$_3$, and imidazole methyl ethyl trichloroaluminate, EtMeIMCl/AlCl$_3$) were used, reaching desulfurization levels above 90% when employing acid ILs prepared with tetrachloroaluminate (0.35:0.65) in excess with respect to the IL precursor. Diesel with 500 ppm of DBT in dodecane (model) along with pre-desulfurized diesel containing 375 ppm of sulfur was employed.

Likewise, Huang *et al.* carried out a similar study featuring an IL with the anion CuCl$_3^-$, reporting sulfur removal of just 23% for a mass ratio of 5:1 (model gasoline:IL) and contact time equal to 30 min [44].

In 2010, our research group published the article: Ionic Liquids Screening for Desulfurization of Natural Gasoline by Liquid-Liquid Extraction. N. Likhanova, D. Guzmán, E. Flores, J. Palomeque, M. A. Domínguez, P. García, Rafael Martínez-Palou. *Mol. Div.* 2010, 14, 777-789. In this context, it should be kept in mind that gasoline is a highly volatile liquid that is recovered cryogenically from natural gas, for it features vapor pressure between the condensate and liquefied petroleum gas (LPG). Natural gasoline is mainly used as a mixture agent for engine gasoline and for refineries and ethylene demand. It is a mixture of liquid hydrocarbons recovered at normal pressure and temperature and is more volatile and unstable than commercial gasoline; it contains mainly C5 and C6 hydrocarbons, aromatic compounds and some sulfur compounds in ppm, especially low molecular weight mercaptan molecules, typically thiols [45].

Commonly, this type of gasoline has a very low content of sulfur compounds, however, in Mexico, it is usual that this gasoline presents sulfur levels above 200 ppm and, in this context, ILs can be an alternative to reduce sulfur below 20 ppm through liquid-liquid extraction.

In the previously mentioned article, a study on the performance of 75 ILs as sequestering agents of sulfur compounds in natural gasoline was carried out. Desulfurization of real natural gasoline was performed by the liquid-liquid extraction method at ambient temperature and atmospheric pressure. The evaluated ILs were synthesized and characterized by spectroscopic techniques under conventional synthesis conditions, employing microwaves and sonochemical methods. The effect of the molecular structure of the ILs on the desulfurization efficiency of natural gasoline with high sulfur content was studied. The analysis of results revealed that the anion type played a more important role than the cation in the desulfurization process. The ILs with halides, aluminates and ferrates displayed the highest efficiency in removing sulfur, and it was higher when a metal salt in excess was present with a ratio of at least 1: 1.3 during the IL synthesis containing the metal anion obtained from its precursors, *i.e.*, the halogenated IL. A recovery methodology of the ILs with metal anion was proposed in the work, which allowed their use in several extraction cycles. This methodology results very interesting if the industrial application of this protocol is considered; this invention has been protected with the following IMP patents:

- R. Martínez-Palou, N. Likhanova, E. A. Flores, D. Guzmán. Líquidos iónicos libre de halógenos en la desulfuración de naftas ligeras y su recuperación. Mexican Patent MX/E/2008/056453 [46].
- D. Guzmán, N. Likhanova, R. Martínez-Palou, E. A. Flores. Proceso para la recuperación de líquidos iónicos agotados en la desulfuración extractiva de naftas. Mexican Patent MX/E/2010/014597 [47].
- R. Martínez-Palou, N. Likhanova, E. A. Flores, D. Guzmán. Halogen-free Ionic Liquids in Naphtha Desulfurization and their Recuperation. German Patent No. 10 2009 039 176.2 [48].
- N. V. Likhanova, R. Martínez-Palou. Desulfurization of Hydrocarbons by Ionic Liquids and preparation of ionic liquids. US Pat. No. 8,821,716 B2 [49].
- D. Guzmán, N.V. Likhanova, R. Martínez-Palou, J. F. Palomeque. Process to recover exhausted ionic liquids used in extractive desulfurization of naphtha. American Patent US/2011/0215052A1 [50].

In this work, an explanation for the desulfurization efficiency displayed by the Lewis-acid-type ILs containing metal anions was proposed by establishing a ratio between the atomic radius and the ionic charge of the anion metal; in this way, the highly efficient performance displayed by the ILs containing ferrates and aluminates as anions was justified (Fig. **2.5**).

In order to find a satisfactory explanation for the performance of ILs containing ferrate-type anions in the desulfurization of hydrocarbons, a study of the type Density Functional Theory (DFT) was carried out, where the interaction between ethanethiol, one of the sulfur compounds detected at high concentration in natural gasoline, and an IL featuring tetrachloroferrate as anion was analyzed in the following paper: DFT Study of the Interaction between Ethanethiol and Fe-containing Ionic Liquids for Desulfurization of Natural Gasoline. J. M. Martínez-Magadán, R. Oviedo-Roa, P. García, R. Martínez-Palou. *Fuel Proc. Technol.* 2012, 97, 24-29 [51].

In this work, it was found that the excess of $FeCl_3$ in the IL synthesis favors the formation of the binuclear anion $Fe_2Cl_7^-$ whose Fe-Cl-Fe bonds are longer and weaker than the Fe-Cl bonds in mononuclear anions such as $FeCl_4^-$; the first bonds are activated for ethanethiol chemisorption. The molecular border orbitals and Mulliken atomic charges have revealed that the high desulfurization yield could be due to a mechanism like the one described by the Dewar-Chatt-Duncanson model of electron donation-backdonation between the sulfur in ethanethiol and the centers of transition metals in the $Fe_2Cl_7^-$ anions, and this mechanism is promoted due to the symmetry affinity between the ethanethiol Highest Occupied Molecular Orbital (HOMO) and atomic orbital t_{2g} in the Fe sites in $Fe_2Cl_7^-$ Lowest Unoccupied Molecular Orbital (LUMO) (Fig. **2.6**).

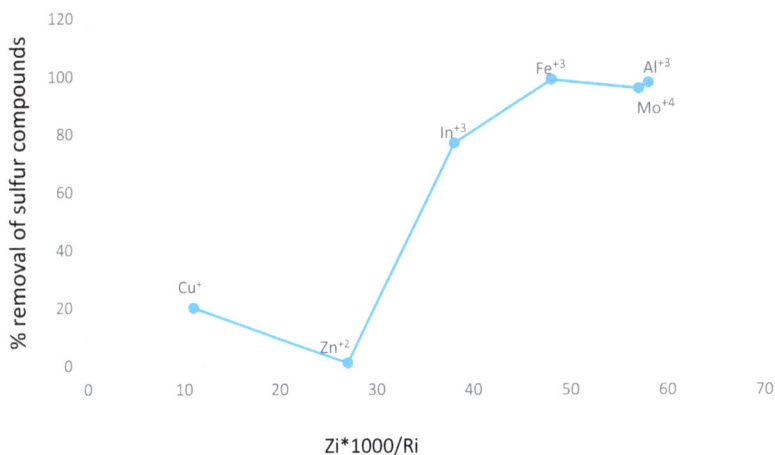

Fig. (2.5). Relationship between the removal efficiency of sulfur compounds and the atomic radius: ionic charge of the anion metal ratio.

In the referred review published by our research group in 2014, a table summarizing some of the works on the desulfurization of hydrocarbons through liquid-liquid extraction employing ILs that appeared between 2004 and 2014 was displayed. From 2015 to date, the harvest of published works on this topic shows that a wide variety of ILs have continued to be evaluated, which confirms the versatility of ILs for this application [52 - 65]. In this context, extractive desulfurization employing ILs has been reviewed and compiled by other authors [66, 67].

Oxidative Desulfurization

Another desulfurization alternative employing ILs is the one known as oxidative desulfurization, to which a section was devoted in the mentioned 2014 review and where a table summarizing some of the most relevant and recent works on this topic was also shown [68, 69].

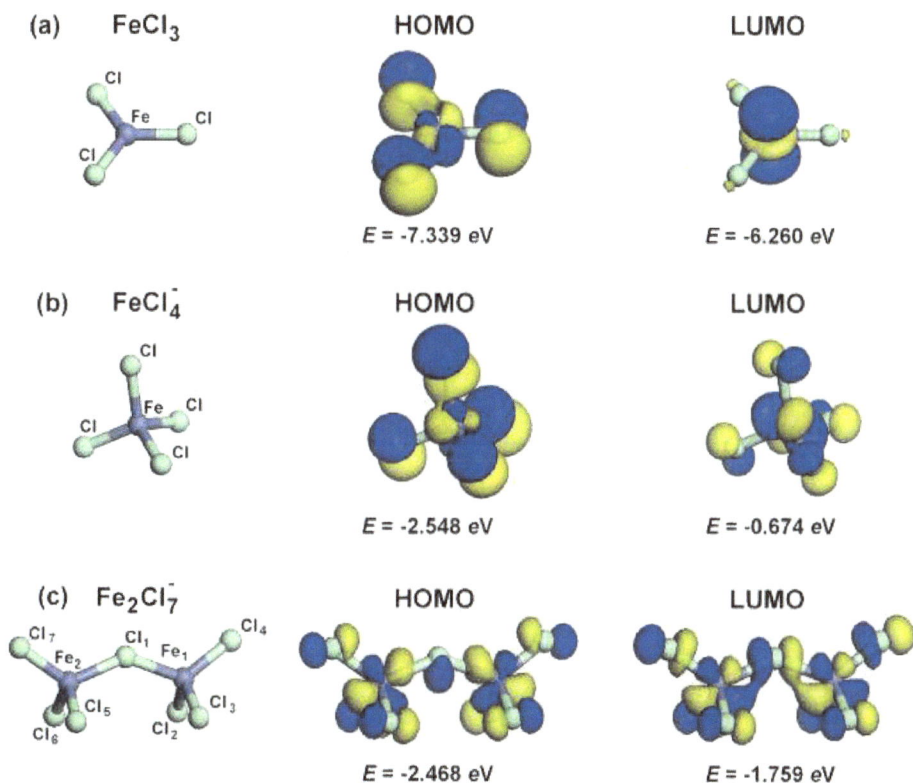

Fig. (2.6). Border electronic states of the highest occupied molecular orbital (HOMO) and the lowest unoccupied molecular orbital (LUMO) for the following species: (a) $FeCl_3$, (b) $FeCl_4^-$ and (c) $Fe_2Cl_7^-$.

With this method, sulfur compounds are previously oxidized to the corresponding sulfoxides and sulfones, which increases the polarity and makes more efficient the extraction through an immiscible solvent, where ILs have turned out to be very efficient candidates (Fig. **2.7**).

In this process, different oxidants such as molecular oxygen [70], hydrogen peroxide [71 - 74], formic acid [75], and acetic acid [76] have been employed.

Other oxidants like HNO_3/NO_2 [77], ozone [78], *tert*-butylhydroperoxide (*t*-BuOOH) [79], and potassium superoxide [80], among others, have been used along with catalysts, in some cases, to make the oxidation process more effective and faster [81 - 82].

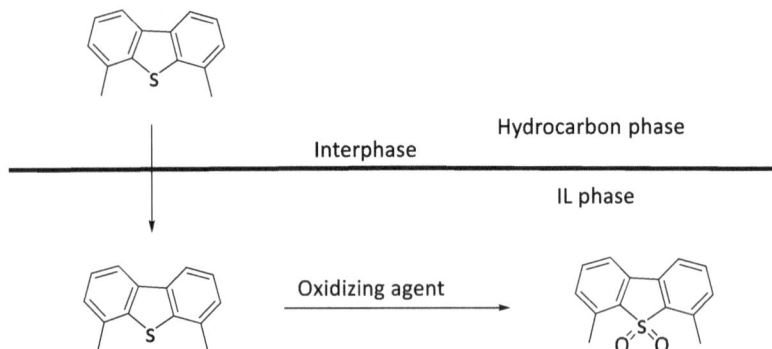

Fig. (2.7). DBT extraction using ILs through oxidative desulfurization.

The oxidative desulfurization of hydrocarbons continues being a very active research field that has prompted many deep reviews of the topic in recent years [83, 84].

Although the obtained advances in the extraction methodologies of sulfur compounds employing ILs have been significant, from a practical point of view, the industrial implementation of this technology has had little progress, for from our standpoint, it has the following limitations:

- Production difficulties and high industrial cost of ILs capable of satisfying the demand for the products required for gasoline desulfurization.
- Even when uncountable ILs with extractive properties of sulfur compounds from hydrocarbon mixtures have been reported, until now, only Lewis-acid-type ILs containing MX_n^- type metal anions, where M is essentially Al and Fe, can remove sulfur compounds from real samples with efficiencies above 90% through a single extraction stage employing hydrocarbon/IL (w/w) proportions not higher than 10:1. To our mind, these ILs would be the only technically and financially viable candidates to be applied at industrial level, however, due to the reactivity of these compounds, they suffer a fast decomposition when reacting with Lewis bases, which is the case of water and sulfur compounds; for this reason, their regeneration and reuse for several extraction cycles is complicated or not possible.

Other aspects that do not make viable the development of this technology are described below:

- Volumetric hydrocarbon losses occur due to three fundamental factors: 1) difficulties in the exhaustive separation of both phases at the interphase zone; 2) unlike the HDS technology, where the reduction reaction produces hydrocarbons from sulfur compounds, in this case, the sulfur compounds are separated, losing part of the hydrocarbon fraction that is integrated to gasoline in the HDS process; and 3) due to the dissolution of a small fraction of hydrocarbon compounds in some ILs.
- The removal methodology of sulfur compounds by extraction with ILs cannot replace HDS, because the desulfurization of crude oil is not really viable because of the high viscosity and la arge amount of crude impurities, which make the extraction process highly inefficient and cause further regeneration of the ILs, for they are not selective. Anyway, EDS using ILs could be a post-HDS alternative for reducing the sulfur content in naphthas to values below 20 ppm, which is difficult to achieve with the current HDS technologies.
- In the case of oxidative desulfurization, the process becomes expensive with respect to the EDS technology, because an oxidizing agent and a catalyst, which are normally expensive and difficult to synthesize, are required, although this limitation can be compensated by employing cheaper and capable-of-beig-regenerated ILs.
- The constant possibility of undesirable reactions taking place in the case of oxidative desulfurization.

Notwithstanding, the desulfurization of hydrocarbons using ILs remains an active research field for the search of alternatives capable of solving the mentioned limitations, for they offer attractive advantages such as hydrogen-free work under mild reaction conditions (low temperature and pressure), better efficiency than the one displayed by HDS removing aromatic sulfur compounds, very simplified industrial implementation with respect to HDS and no expensive, noble-metal-based catalysts are required.

Application of ILs in the Denitrogenation of Oil-derived Fuels

Most heavy crude oils in the world, and more specifically the Mexican crude oil types, are characterized by having a high content of nitrogen compounds, which in addition to generating toxic gases, are important inhibitors of the HDS reaction; for this reason, the previous removal of nitrogen compounds contributes to reach the required sulfur levels under milder conditions and extend the lifespan of catalysts.

Currently, a more promising alternative for the extraction of ILs would be the removal of nitrogen compounds from the feedstocks that will be used in the HDS process to generate UBA loads. It is known that the removal of nitrogen compounds makes more efficient the HDS process. The advantage of using ILs to this end is that unlike what occurs with sulfur compounds, first generation ILs, *i.e.* those obtained through a single synthesis stage from alkyl halides, thus displaying halide anions [85, 86], have shown to be effective in removing nitrogen compounds, in addition to being cheaper (they are obtained through a single stage), more chemically and thermally stable and can be regenerated (using just water) and reused for several reaction cycles without sensible activity loss; furthermore, besides removing nitrogen compounds, they remove partially sulfur compounds present in the feedstocks.

The nitrogen compounds in oil derivatives can be divided into two types: heterocyclic and non-heterocyclic. The last ones include aliphatic amines and anilines whereas the first category can be subdivided into two families: basic and neutral [87]. The most common family in the basic nitrogen compounds are the six-member pyridinic rings whereas the neutral nitrogen compounds are those containing five-member pyrrolic rings. Table **2.2** shows the different nitrogen compounds present in fossil fuels. The basic nitrogen compounds are those that largely inhibit the HDS process [88].

Both families of heterocyclic nitrogen compounds have different electron configurations and then, they interact differently with the catalyst surface. In the five-member-nitrogen hetero aromatic compounds, the nitrogen extra electron pair, which usually confers basicity to the nitrogen compounds, is involved in the ring electron cloud and then, it is not available to interact with acids. Since nitrogen in the pyrrole aromatic ring works as an electron source, a pyrrolic ring is relatively rich in electrons (excessive π) in comparison with a benzene ring. It can be expected that the initial contact of these nitrogen heterocycles with the catalyst surface is mainly associated with the ring high electron density and not with nitrogen.

In contrast, because the nitrogen electron pair in the six-member heteroaromatic rings is not involved in the π electron cloud, it is then available to be shared with acids. These compounds are strong bases. Because of the attractive nature of the nitrogen atom electrons in the pyridinic ring, the six-member heteroaromatic rings are relatively deficient in electrons (π deficient) with respect to their benzenoid counterpart. It is likely that these compounds use nitrogen preferably to carry out the initial contact with the catalyst surface if this heteroatom is not sterically hindered.

Table 2.2. Main nitrogen compounds present in fossil fuels.

Type	Compound
Six-member-heterocyclic rings	Pyridine
	Quinoline
	Tetrahydroquinoline
	Acridine
Five-member heterocyclic rings	Pyrrole
	Indole
	Indoline
	Carbazole
Non-heterocyclic	Aniline

The basicity of these nitrogen compounds allows the interaction with the acid active sites located on the catalyst surface through either of the following forms: they can accept protons from the surface (Brønsted acidity) or can donate unpaired electron pairs to deficient-electron sites on the same surface (Lewis acidity) [89].

In 2012, IMP researchers published the article entitled: Parallel Microwave-assisted Synthesis and Screening of Ionic Liquids for Denitrogenation of Straight-Run Diesel Feed by liquid-liquid extraction. M. A. Cerón, D. Guzmán-Lucero, J. F. Palomeque, R. Martínez-Palou. Comb. Chem. High Throughput Screen. 2012, 15, 427-432. In this work, fifty-six ILs were synthesized, characterized, and evaluated as extraction agents of nitrogen compounds present in a diesel feedstock before being submitted to the HDS process; the aim was to remove simultaneously both nitrogen compounds and part of the sulfur compounds present in the fuel. The results of this study showed that some ILs, in particular those containing transition metals and halogens as part of the anion, could remove a high content of nitrogen and sulfur compounds through a liquid-liquid extraction process [90].

It was found that several IL types displayed good performance in the selective extraction of nitrogen compounds by means of the liquid-liquid extraction. Particularly, halogenated ILs exhibited good performance, and since they were first generation ILs (synthesized at a single stage), their production was relatively cheap and showed high selectivity in the extraction of nitrogen compounds in addition to their capacity to be regenerated and recycled. Also, the ILs containing ferrate- and aluminate-type anions resulted very effective for this purpose, *e.g.* the IL with the cation 1-methyl-3-octylimidazolium and the anion tetrachloroferrate ([OMIM]ClFeCl$_3$ (1:1.5)) diminished the original content in real diesel (466 ppm of N) in 95%, employing an IL/diesel ratio of 1/5 for 10 min of stirring at 60°C.

The described results were protected through the following patents:

- Desnitrogenación de hidrocarburos mediante un proceso de extracción líquido-líquido empleando líquidos iónicos. R. Martínez-Palou, D. Guzmán, J.F. Palomeque. Patente Mexicana MX/E/2011/051368.
- Denitrogenation of Hydrocarbons by liquid-liquid extraction using ionic liquids. R. Martínez-Palou, D. Guzmán, J. F. Palomeque. Patente U.S. 9,157,034B2.
- Denitrogenation of Hydrocarbons by liquid-liquid extraction using ionic liquids. R. Martínez-Palou, D. Guzmán, J. F. Palomeque. Patente Canadiense 2,783,754.

IMP researchers also studied the effect exerted by ILs on the denitrogenation of feedstocks destined to the production of diesel in the paper by Laredo, G. C.; Likhanova, N. V.; Lijanova, I. V.; Rodriguez-Heredia, B.; Castillo, J. J.; and Perez-Romo, P. entitled Synthesis of ionic liquids and their use for extracting nitrogen compounds from gas oil feeds towards diesel fuel production. Fuel Proc. Technol. 2015, 130, 38-45 [91]. In this work, seventeen ILs with imidazole- and quaternary ammonium-type cations were evaluated. To assess the denitrogenation, a model mixture consisting of quinoline, indole and carbazole in hexadecane/toluene was employed in this study. The denitrogenation tests were carried out at 30°C with a feedstock/IL ratio of 20/1. Also, a speciation of nitrogen compounds was performed, observing that, in general, the evaluated ILs displayed higher removal levels of indole and carbazole than of quinoline. In addition, when the number of carbon atoms increased in the counterpart (quaternary ammonium carboxylate), the removal of carbazole also augmented. Additional experiments using model mixtures containing benzothiophene and benzothiophene and aniline showed that none of the compounds in the mixture affected the IL solubility under the same experimental conditions.

The experiments carried out using a batch configuration and gasoil confirmed that only triethylammonium butyrate and triethylammonium acetate displayed good performance removing nitrogen and suitable chemical stability under the studied experimental conditions. Then, these compounds were selected for a dynamic study, which showed that the extraction of organic nitrogen was achieved at 30% after flowing 200 and 170 mL per gram of material before requiring a regeneration process.

The triethylammonium derivatives with acetate and butyrate chains were chosen as potential candidates for a large-scale application due to their nitrogen removal capacity, in addition to their simple synthetic procedures and relatively low synthesis costs with respect to their imidazole-derivative analogues.

IMP researchers also studied the denitrogenation of intermediate distillates employing adsorbent inorganic materials [92].

The denitrogenation of hydrocarbons through EDN has been widely researched in the last years [93 - 101] and was reviewed in 2016 by Abro *et al*. [102] and by Prado *et al*. [103] In the last work, it was concluded that EDN using ILs is an independent process of temperature and that the equilibrium time is not an obstacle when low viscosity ILs such as [EMIM]N(CN)$_2$ and [BMIM]N(CN)$_2$ are employed; these compounds are excellent candidates for the denitrogenation of hydrocarbons, for they can be regenerated for various extraction cycles without evident loss of their extraction capacity.

Application of ILs in the Removal of Fluorinated Compounds from Alkylation Gasolines

Another alternative application of the ILs through the liquid-liquid extraction process is the removal of fluorinated compounds present in alkylation gasoline when hydrofluoric acid is employed to produce it. The alkylation process will be discussed in more details in Chapter 9 of this book.

In the petrochemical industry, the natural crude compounds are separated traditionally by distillation. However, this process does not change the relative proportions of the individual components, where some of them have higher commercial values than others.

The alkylation process implies better use of the oil fractions; normally, the low boiling point fractions such as propylene and butylene react with isobutane to produce high octane products. Such products originate fuel that is burned efficiently, extending the engine lifespan, and reducing the emissions. By increasing the fuel octane number, alkylates prevent the use of lead, which is harmful for the environment, without affecting the engine efficiency.

Alkylation is a chemical synthesis process that consists of the reaction between light olefins and saturated hydrocarbons, producing branched-chain-saturated hydrocarbons, which are also known as high-octane-index gasoline. This is a catalytic process that requires acid-nature catalysts such as hydrofluoric acid (HF) or sulfuric acid (H_2SO_4). Fig. (**2.8**) shows the general diagram of a process to obtain alkylation gasoline [104].

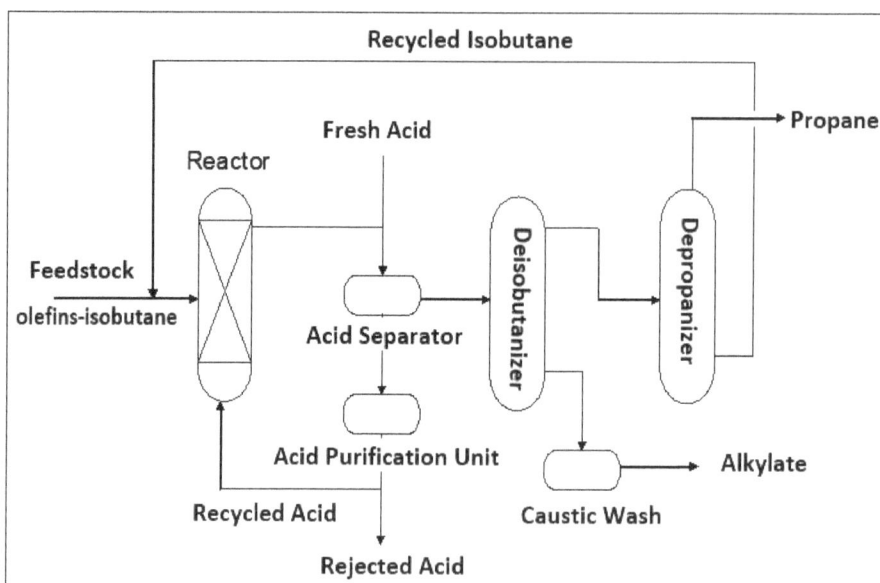

Fig. (2.8). General alkylation process with HF.

In the oil industry, this term is referred specifically to the catalytic process devoted to the alkylation of isobutane with various light olefins (ethylene, propylene, pentenes and butene dimers) to produce highly branched paraffins. Then, *e.g.* 2,2,4-trimethylpentane (isooctane) is formed as main product from either isobutene or isobutane (Fig. **2.9**).

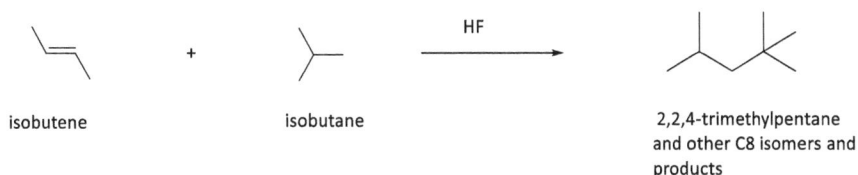

Fig. (2.9). Reaction to obtain 2,2,4-trimethylpentane and other C_8 isomers.

Although in the alkylation process there are controls for trapping the fluorinated compounds and HF produced during the process, alkylation gasoline can be contaminated with these compounds that are highly toxic, and for this reason, our research team was focused on studying the use of ILs in the extraction of this type of pollutants [105].

Some technologies propose the removal of HF and RF from the gaseous hydrocarbon currents (like C3 and C4) at the end of the process by means of adsorbent materials such as zeolites, zeolitas [106 - 109], heteropolyacids [110], or silica/Nafion nanocomposites, but in practice, KOH diluted in water is commonly employed [101], however, most of these materials either undergo fast deactivation or are not suitable for industrial applications.

In 2019, our research team published the work: Experimental and Theoretical Assessment of Fluorinated Compounds Removal from Organic Media using Ionic Liquids. Alma D. Miranda, M. Gallo, J. M. Domínguez, Rafael Martínez-Palou. *J. Mol. Liq.* 2019, 276, 779-793. This manuscript describes the potential use of three ILs to remove efficiently fluorinated compounds from alkylation gasoline obtained from a production process in Mexico. As far as we know, this was the first time that the use of ILs for this purpose was proposed [112].

The ILs that were synthesized and evaluated in the extraction of HF and fluorinated organic compounds were 1-butyl-3-methyilimidazolium, [BMIM][Br], 1-butyl-3-methylimidazolium trifluoroacetate, [BMIM][CF_3COO], and 1-butyl-3-methylimidazolium bis (trifluoromethanesulfate) [BMIM][NTf_2]; the evaluation of these ILs in the adsorption of HF and RF species from organic medium (AG) was carried out experimentally.

According to the experimental results, the ILs resulted to be good extraction agents of fluorinated compounds and especially of HF. The two ILs containing fluorinated anions displayed efficiency that was better than that of [BMIM][Br].

Also, the molecular interactions between HF and the ILs in the organic media were studied through hybrid theoretical calculations "QM (quantum mechanics)/MM (molecular mechanics)- Molecular Dynamics (MD)". These QM/MM-MD simulations allowed to establish a series of configurations of the HF-IL system equidistant in time to evaluate the interaction energies at the QM theoretical level between HF and the ILs; also, the mapping of non-covalent interactions was performed. According to the theoretical calculations, it seems that the anion capacity in the ILs to form hydrogen bridge bonds with HF is the driving force in the HF extraction. Previously, ILs had been proposed as possible catalysts of the alkylation reaction [112 - 113]. Currently, the IMP is studying other alternatives to synthesize more stable, IL-based catalysts employing poly-ILs and supported-ILs; the results obtained from these studies will be published soon.

The applications of ILs in desulfurization, denitrogenation and as catalysts to the alkylation proceses were reviewed recently by Salah *et al* [114].

CONCLUDING REMARKS

In this chapter, the application of ILs for removing pollutants from refinery feedstocks, like remotion of sulfur-, nitrogen- and fluor-containing compounds, aromatics, naphthenic acids and asphaltenes is reviewed. Critical considerations for the applications of these technologies in petroleum industry are discussed, demonstrating the challenges, prospects, and limitations of ILs in this research area. The industrial application related to the removal of contaminants still needs to go a long way for its implementation at the industrial level, fundamentally due to the costs of the process related to the limited regenerability and reuse in successive cycles of the ILs; for example, ILs containing anions based on Lewis acids show excellent performance in the removal of sulfur and nitrogen compounds, particularly those whose metathesis is carried out with aluminum or iron salts; however, these ILs are sensitive to moisture, which makes their large-scale application and reuse difficult.

REFERENCES

[1]   Kirk-Orthmer Encyclopedia of Chemical Technology. Wiley, U.K, **2006**. 20.

[2]   Wei, S.; Wang, F.; Dan, M.; Yu, S.; Zhou, Y. Vanadium (V) and Niobium (Nb) as the most promising co-catalysts for hydrogen sulfide splitting screened out from 3d and 4d transition metal single atoms. *Int. J. Hydrogen Energy,* **2020**, *45*(35), 17480-17492.
[http://dx.doi.org/10.1016/j.ijhydene.2020.04.266]

[3]   Vrinat, M.L. The kinetics of the hydrodesulfurization process - a review. *Appl. Catal.,* **1983**, *6*(2), 137-158.
[http://dx.doi.org/10.1016/0166-9834(83)80260-7]

[4]   Rigutto, M.S.; van Veen, R.; Huve, L. Zeolites in Hydrocarbon Processing. In: *Studies in Surface Science and Catalysis*; Čejka, J.; van Bekkum, H.; Corma, A.; Schüth, F., Eds.; Elsevier: London, **2007**.

[5]   Saleh, T.A. Global trends in technologies and nanomaterials for removal of sulfur organic compounds: Clean energy and green environment. *J. Mol. Liq.,* **2022**, *359*, 119340.
[http://dx.doi.org/10.1016/j.molliq.2022.119340]

[6]   Kaufmann, T.G.; Kaldor, A.; Stuntz, G.F.; Kerby, M.C.; Ansell, L.L. Catalysis science and technology for cleaner transportation fuels. *Catal. Today,* **2000**, *62*(1), 77-90.
[http://dx.doi.org/10.1016/S0920-5861(00)00410-7]

[7]   Marcilly, C. Evolution of refining and petrochemicals. *Proceedings of the 13th International Zeolite Conference,* **2001**, pp. 37-60.
[http://dx.doi.org/10.1016/S0167-2991(01)81185-X]

[8]   Song, C.; Ma, X. New design approaches to ultra-clean diesel fuels by deep desulfurization and deep dearomatization. *Appl. Catal. B,* **2003**, *41*(1-2), 207-238.
[http://dx.doi.org/10.1016/S0926-3373(02)00212-6]

[9]   Velázquez, H.D.; Cerón-Camacho, R.; Mosqueira-Mondragón, M.L.; Hernández-Cortez, J.G.; Montoya de la Fuente, J.A.; Hernández-Pichardo, M.L.; Beltrán-Oviedo, T.A.; Martínez-Palou, R. Recent progress on catalyst technologies for high quality gasoline production. *Catal. Rev., Sci. Eng.,* **2022**, 1-221.
[http://dx.doi.org/10.1080/01614940.2021.2003084]

[10]   Miller, J.T.; Reagan, W.J.; Kaduk, J.A.; Marshall, C.L.; Kropf, A.J. Selective Hydrodesulfurization of FCC Naphtha with Supported MoS2 Catalysts: The Role of Cobalt. *J. Catal.,* **2000**, *193*(1), 123-131.
[http://dx.doi.org/10.1006/jcat.2000.2873]

[11]   Corma, A.; Martínez, C.; Ketley, G.; Blair, G. On the mechanism of sulfur removal during catalytic cracking. *Appl. Catal. A Gen.,* **2001**, *208*(1-2), 135-152.
[http://dx.doi.org/10.1016/S0926-860X(00)00693-1]

[12]   Yin, C.; Zhu, G.; Xia, D. Determination of organic sulfur compounds in naphtha. Part II. Identification and quantitative analysis of thiophenes in FCC and RFCC naphthas. *Am Chem. Soc. Prepr. Div. Pet. Chem,* **2002**, *47*, 398-401.

[13]   Corbett, R.A. Tougher diesel specs could force major refining industry expenditures. *Oil Gas J.,* **1987**, *85*, 56-59.

[14]   Swaim, E.J. Major growth in coke production takes place. *Oil Gas J.,* **1991**, *89*, 100-102.

[15]   Brunet, S.; Mey, D.; Pérot, G.; Bouchy, C.; Diehl, F. On the hydrodesulfurization of FCC gasoline: a review. *Appl. Catal. A Gen.,* **2005**, *278*(2), 143-172.
[http://dx.doi.org/10.1016/j.apcata.2004.10.012]

[16]   Zhong, Q.; Shen, H.; Yun, X.; Chen, Y.; Ren, Y.; Xu, H.; Shen, G.; Du, W.; Meng, J.; Li, W.; Ma, J.; Tao, S. Global Sulfur Dioxide Emissions and the Driving Forces. *Environ. Sci. Technol.,* **2020**, *54*(11), 6508-6517.
[http://dx.doi.org/10.1021/acs.est.9b07696] [PMID: 32379431]

[17]   He, J. Pollution haven hypothesis and environmental impacts of foreign direct investment: The case of industrial emission of sulfur dioxide (SO_2) in Chinese provinces. *Ecol. Econ.,* **2006**, *60*(1), 228-245.
[http://dx.doi.org/10.1016/j.ecolecon.2005.12.008]

[18]   Zhang, Q.; Nakatani, J.; Shan, Y.; Moriguchi, Y. Inter-regional spillover of China's sulfur dioxide (SO_2) pollution across the supply chains. *J. Clean. Prod.,* **2019**, *207*, 418-431.
[http://dx.doi.org/10.1016/j.jclepro.2018.09.259]

[19]   Martínez-Palou, R.; Flores, P. Perspectives of Ionic Liquids for Clean Oilfield Technologies. In: *Ionic Liquids. Theory, Properties, New Approaches*; Kokorin, A., Ed.; INTECH, **2011**; pp. 567-630.
[http://dx.doi.org/10.5772/14529]

[20]   Luque, R.; Martínez-Palou, R. Applications of Ionic liquids for Removing Pollutants from Refinery Feedstocks: A review. *Environm. Energy Sci,* **2014**, *7*, 2414-2447.

[http://dx.doi.org/10.1039/C3EE43837F]

[21]   Xu, X.; Chen, Y. Air emissions from the oil and natural gas industry. *Int. J. Environ. Stud.,* **2016**, *73*(3), 422-436.
[http://dx.doi.org/10.1080/00207233.2016.1165483]

[22]   Nasiritousi, N. Fossil fuel emitters and climate change: unpacking the governance activities of large oil and gas companies. *Env. Polit.,* **2017**, *26*(4), 621-647.
[http://dx.doi.org/10.1080/09644016.2017.1320832]

[23]   MacCracken, M.C. The Increasing Pace of Climate Change. *Strateg. Plann. Energy Environ.,* **2009**, *28*(3), 8-25.
[http://dx.doi.org/10.1080/10485230909509197]

[24]   Mexicanas, N.O.; de Fomento, S.; Ambiental, N. **2007**. http://www.semarnat.gob.mx/leyesynormas/normasoficialesmexicanasvigentes/pages/inicio/ (in Spanish)

[25]   Laws, R. Environmental Protection Agency. **2002**.http://www.epa.gov/epahome/lawregs/

[26]   Calidad de las gasolinas y de los combustibles Diesel: azufre y plomo, Unión Europea. http://europa.eu/scadplus/leg/es/lvb/l28077.htm (in Spanish)

[27]   Tanimu, A.; Alhooshani, K. Advanced Hydrodesulfurization Catalysts: A Review of Design and Synthesis. *Energy Fuels,* **2019**, *33*(4), 2810-2838.
[http://dx.doi.org/10.1021/acs.energyfuels.9b00354]

[28]   Hu, S.; Luo, G.; Shima, T.; Luo, Y.; Hou, Z. Hydrodenitrogenation of pyridines and quinolines at a multinuclear titanium hydride framework. *Nat. Commun.,* **2017**, *8*(1), 1866.
[http://dx.doi.org/10.1038/s41467-017-01607-z] [PMID: 29192198]

[29]   Islam, M.R.; Chhetri, A.B.; Khan, M.M. *Greening of Petroleum Operations: The Science of Sustainable Energy Production*; Wiley-Scrivener: Massachusetts, **2010**.
[http://dx.doi.org/10.1002/9780470922378]

[30]   An, G.J.; Zhou, T.N.; Chai, Y.M.; Zhang, J.C.; Liu, Y.Q.; Liu, C.G. Nonhydrodesulfurization Technologies of Light Oil. *Huaxue Jinzhan,* **2007**, *19*, 1331-1344.

[31]   Zaczepinski, S. Exxon Diesel Oil Deep Desulfurization (DODD). In: *Handbook of Petroleum Refining Processes*; Meyer, R.A., Ed.; McGraw-Hill: New York, **1996**.

[32]   Kabe, T.; Ishihara, A.; Qian, W. *Hydrodesulfurization and Hydrodenitrogenation: Chemistry and Engineering*; Willey-VCH: Weinheim, **1999**.

[33]   Ferrari, M.; Maggi, R.; Delmon, B.; Grange, P. Influences of the Hydrogen Sulfide Partial Pressure and of a Nitrogen Compound on the Hydrodeoxygenation Activity of a CoMo/Carbon Catalyst. *J. Catal.,* **2001**, *198*(1), 47-55.
[http://dx.doi.org/10.1006/jcat.2000.3103]

[34]   Caeiro, G.; Costa, A.F.; Cerqueira, H.S.; Magnoux, P.; Lópes, J.M.; Matias, P.; Ribeiro, F.R. Nitrogen poisoning effect on the catalytic cracking of gasoil. *Appl. Catal. A Gen.,* **2007**, *320*, 8-15.
[http://dx.doi.org/10.1016/j.apcata.2006.11.031]

[35]   Kulkarni, P.S.; Afonso, C.A.M. Deep desulfurization of diesel fuel using ionic liquids: current status and future challenges. *Green Chem.,* **2010**, *12*(7), 1139-1149.
[http://dx.doi.org/10.1039/c002113j]

[36]   Srivastava, V.C. An evaluation of desulfurization technologies for sulfur removal from liquid fuels. *RCS Adv,* **2012**, *2*, 759-783.
[http://dx.doi.org/10.1039/C1RA00309G]

[37]   Huang, C.; Chen, B.; Zhang, J.; Liu, Z.; Li, Y. Desulfurization of Gasoline by Extraction with New Ionic Liquids. *Energy Fuels,* **2004**, *18*(6), 1862-1864.
[http://dx.doi.org/10.1039/C1RA00309G]

[38]   Babich, I.; Moulijin, J.A. Science and technology of novel processes for deep desulfurization of oil refinery streams: a review. *Fuel,* **2003**, *82*(6), 607-631.
[http://dx.doi.org/10.1016/S0016-2361(02)00324-1]

[39]   Song, C. An overview of new approaches to deep desulfurization for ultra-clean gasoline, diesel fuel and jet fuel. *Catal. Today,* **2003**, *86*(1-4), 211-263.
[http://dx.doi.org/10.1016/S0920-5861(03)00412-7]

[40]   Liu, F.; Yu, J.; Qazi, A.B.; Zhang, L.; Liu, X. Metal-based ionic liquids in oxidative desulfurization: a critical review. *Environ. Sci. Technol.,* **2021**, *55*(3), 1419-1435.
[http://dx.doi.org/10.1021/acs.est.0c05855] [PMID: 33433212]

[41]   Ito, E.; van Veen, J.A.R. On novel processes for removing sulphur from refinery streams. *Catal. Today,* **2006**, *116*(4), 446-460.
[http://dx.doi.org/10.1016/j.cattod.2006.06.040]

[42]   Stanislaus, A.; Marafi, A.; Rana, M.S. Recent advances in the science and technology of ultra low sulfur diesel (ULSD) production. *Catal. Today,* **2010**, *153*(1-2), 1-68.
[http://dx.doi.org/10.1016/j.cattod.2010.05.011]

[43]   Bösmann, A.; Datsevich, L.; Jess, A.; Lauter, A.; Schmitz, C.; Wasserscheid, P. Deep desulfurization of diesel fuel by extraction with ionic liquids. *Chem. Commun. (Camb.),* **2001**, (23), 2494-2495.
[http://dx.doi.org/10.1039/b108411a] [PMID: 12240031]

[44]   Dharaskar, S.A.; Wasewar, K.L.; Varma, M.N.; Shende, D.Z. Imidazolium ionic liquid as energy efficient solvent for desulfurization of liquid fuel. *Separ. Purif. Tech.,* **2015**, *155*, 101-109.
[http://dx.doi.org/10.1016/j.seppur.2015.05.032]

[45]   Likhanova, N.V.; Guzmán-Lucero, D.; Flores, E.A.; García, P.; Domínguez-Aguilar, M.A.; Palomeque, J.; Martínez-Palou, R. Ionic liquids screening for desulfurization of natural gasoline by liquid–liquid extraction. *Mol. Divers.,* **2010**, *14*(4), 777-787.
[http://dx.doi.org/10.1007/s11030-009-9217-x] [PMID: 20091120]

[46]   Martínez-Palou, R.; Likhanova, N.; Flores, E.A.; Guzmán, D. Líquidos iónicos libre de halógenos en la desulfuración de naftas ligeras y su recuperación. *Mexican Pat. MX/E/2008/056453 (in Spanish),* **2008**.

[47]   Guzmán, D.; Likhanova, N.; Martínez-Palou, R.; Flores, E.A. Proceso para la recuperación de líquidos iónicos agotados en la desulfuración extractiva de naftas. *Mexican Pat. MX/E/2010/014597 (in Spanish),* **2010**.

[48]   Martínez-Palou, R.; Likhanova, N.; Flores, E.A.; Guzmán, D. Halogen-free Ionic Liquids in Naphtha Desulfurization and their Recuperation. *German Pat. 10 2009 039 176.2,* **2009**.

[49]   Likhanova, N.V.; Martínez-Palou, R.; Palomeque, J.F. Desulfurization of Hydrocarbons by Ionic Liquids and preparation of ionic liquids. U.S. Pat. No. 8,821,716 B2.

[50]   Guzmán, D.; Likhanova, N.V.; Martínez-Palou, R.; Palomeque, J.F. Process to recover exhausted ionic liquids used in extractive desulfurization of naphtha. U.S. Pat. 2011/0215052A1.

[51]   Martínez-Magadán, J.M.; Oviedo-Roa, R.; García, P.; Martínez-Palou, R. DFT study of the interaction between ethanethiol and Fe-containing ionic liquids for desulfuration of natural gasoline. *Fuel Process. Technol.,* **2012**, *97*, 24-29.
[http://dx.doi.org/10.1016/j.fuproc.2012.01.007]

[52]   Enayati, M.; Faghihian, H. N-butyl-pyridinium tetrafluoroborate as a highly efficient ionic liquid for removal of dibenzothiophene from organic solutions. *J. Fuel Chem. Technol.,* **2015**, *43*(2), 195-201.
[http://dx.doi.org/10.1016/S1872-5813(15)30003-7]

[53]   Ahmed, O.U.; Mjalli, F.S.; Al-Wahaibi, T.; Al-Wahaibi, Y.; AlNashef, I.M. Optimumn performance of extractive desulfurization of liquid fuels using phosphonium and pyrrolidinium-based ionic liquids. *Ind. Eng. Chem. Res.,* **2015**, *54*(25), 6540-6550.

[http://dx.doi.org/10.1021/acs.iecr.5b01187]

[54]   Jiang, W.; Zhu, W.; Li, H.; Wang, X.; Yin, S.; Chang, Y.; Li, H. Temperature-responsive ionic liquid extraction and separation of the aromatic sulfur compounds. *Fuel,* **2015**, *140*, 590-596.
[http://dx.doi.org/10.1016/j.fuel.2014.09.083]

[55]   Farzin Nejad, N.; Miran Beigi, A.A. Efficient desulfurization of gasoline fuel using ionic liquid extraction as a complementary process to adsorptive desulfurization. *Petrol. Sci.,* **2015**, *12*(2), 330-339.
[http://dx.doi.org/10.1007/s12182-015-0020-2]

[56]   Rogošić, M.; Sander, A.; Kojić, V.; Vuković, J.P. Liquid–liquid equilibria in the ternary and multicomponent systems involving hydrocarbons, thiophene or pyridine and ionic liquid (1-benzyl-3-metylimidazolium bis(trifluorometylsulfonyl)imide). *Fluid Phase Equilib.,* **2016**, *412*, 39-50.
[http://dx.doi.org/10.1016/j.fluid.2015.12.025]

[57]   Moheb-Aleaba, Z.; RezaKhosravi-Nikou, M. Extractive desulfurization of liquid hydrocarbon fuel: Task-specific ionic liquid development and experimental study. *Chem. Eng. Res. Des.,* **2023**, *189*, 234-249.
[http://dx.doi.org/10.1016/j.cherd.2022.11.021]

[58]   Mafi, M.; Dehghani, M.R.; Mokhtarani, B. Novel liquid–liquid equilibrium data for six ternary systems containing IL, hydrocarbon and thiophene at 25 °C. *Fluid Phase Equilib.,* **2016**, *412*, 21-28.
[http://dx.doi.org/10.1016/j.fluid.2015.12.006]

[59]   Safa, M.; Mokhtarani, B.; Mortaheb, H.R. Deep extractive desulfurization of dibenzothiophene with imidazolium or pyridinium-based ionic liquids. *Chem. Eng. Res. Des.,* **2016**, *111*, 323-331.
[http://dx.doi.org/10.1016/j.cherd.2016.04.021]

[60]   Yu, F.; Liu, C.; Yuan, B.; Xie, P.; Xie, C.; Yu, S. Energy-efficient extractive desulfurization of gasoline by polyether-based ionic liquids. *Fuel,* **2016**, *177*, 39-45.
[http://dx.doi.org/10.1016/j.fuel.2016.02.063]

[61]   Bui, T.T.L.; Nguyen, D.D.; Ho, S.V.; Nguyen, B.T.; Uong, H.T.N. Synthesis, characterization and application of some non-halogen ionic liquids as green solvents for deep desulfurization of diesel oil. *Fuel,* **2017**, *191*, 54-61.
[http://dx.doi.org/10.1016/j.fuel.2016.11.044]

[62]   Li, J.; Lei, X.J.; Tang, X.D.; Zhang, X.P.; Wang, Z.Y.; Jiao, S. Acid Dicationic Ionic Liquids as Extractants for Extractive Desulfurization. *Energy Fuels,* **2019**, *33*(5), 4079-4088.
[http://dx.doi.org/10.1021/acs.energyfuels.9b00307]

[63]   Player, L. C.; Chan, B.; Lui, M. Y.; Masters, A. F.; Maschmeyer, T. Toward an Understanding of the Forces Behind Extractive Desulfurization of Fuels with Ionic Liquids.

[64]   Jiang, B.; Yang, H.; Zhang, L.; Zhang, R.; Sun, Y.; Huang, Y. Efficient oxidative desulfurization of diesel fuel using amide-based ionic liquids. *Chem. Eng. J.,* **2016**, *283*, 89-96.
[http://dx.doi.org/10.1016/j.cej.2015.07.070]

[65]   Wang, L.; Jin, G.; Xu, Y. Desulfurization of coal using four ionic liquids with [HSO4]−. *Fuel,* **2019**, *236*, 1181-1190.
[http://dx.doi.org/10.1016/j.fuel.2018.09.082]

[66]   Kianpour, E.; Azizian, S.; Yarie, M.; Zolfigol, M.A.; Bayat, M. A task-specific phosphonium ionic liquid as an efficient extractant for green desulfurization of liquid fuel: An experimental and computational study. *Chem. Eng. J.,* **2016**, *295*, 500-508.
[http://dx.doi.org/10.1016/j.cej.2016.03.072]

[67]   Chen, X.; Guan, Y.; Abdeltawab, A.A.; Al-Deyab, S.S.; Yuan, X.; Wang, C.; Yu, G. Using functional acidic ionic liquids as both extractant and catalyst in oxidative desulfurization of diesel fuel: An investigation of real feedstock. *Fuel,* **2015**, *146*, 6-12.
[http://dx.doi.org/10.1016/j.fuel.2014.12.091]

[68]   Abro, R.; Abdeltawab, A.A.; Al-Deyab, S.S.; Yu, G.; Qazi, A.B.; Gao, S.; Chen, X. A review of extractive desulfurization of fuel oils using ionic liquids. *RSC Advances,* **2014**, *4*(67), 35302-35317.
        [http://dx.doi.org/10.1039/C4RA03478C]

[69]   Ibrahim, M.H.; Hayyan, M.; Hashim, M.A.; Hayyan, A. The role of ionic liquids in desulfurization of fuels: A review. *Renew. Sustain. Energy Rev.,* **2017**, *76*, 1534-1549.
        [http://dx.doi.org/10.1016/j.rser.2016.11.194]

[70]   Lü, H.; Gao, J.; Jiang, Z.; Yang, Y.; Song, B.; Li, C. Oxidative desulfurization of dibenzothiophene with molecular oxygen using emulsion catalysis. *Chem. Commun. (Camb.),* **2007**, (2), 150-152.
        [http://dx.doi.org/10.1039/B610504A] [PMID: 17180229]

[71]   Dehkordi, A.M.; Kiaei, Z.; Sobati, M.A. Oxidative desulfurization of simulated light fuel oil and untreated kerosene. *Fuel Process. Technol.,* **2009**, *90*(3), 435-445.
        [http://dx.doi.org/10.1016/j.fuproc.2008.11.006]

[72]   Gao, H.; Guo, C.; Xing, J.; Zhao, J.; Liu, H. Extraction and oxidative desulfurization of diesel fuel catalyzed by a Brønsted acidic ionic liquid at room temperature. *Green Chem.,* **2010**, *12*(7), 1220-1224.
        [http://dx.doi.org/10.1039/c002108c]

[73]   Zhao, D.; Sun, Z.; Li, F.; Shan, H. Optimization of oxidative desulfurization of dibenzothiophene using acidic ionic liquid as catalytic solvent. *J. Fuel Chem. Technol.,* **2009**, *37*(2), 194-198.
        [http://dx.doi.org/10.1016/S1872-5813(09)60015-3]

[74]   Chen, X.; Guo, H.; Abdeltawab, A.A.; Guan, Y.; Al-Deyab, S.S.; Yu, G.; Yu, L. Brønsted- Lewis acidic ionic liquids and application in oxidative desulfurization of diesel fuel. *Energy Fuels,* **2015**, *29*(5), 2998-3003.
        [http://dx.doi.org/10.1021/acs.energyfuels.5b00172]

[75]   Gao, J.; Wang, S.; Jiang, Z.; Lu, H.; Yang, Y.; Jing, F.; Li, C. Deep desulfurization from fuel oil *via* selective oxidation using an amphiphilic peroxotungsten catalyst assembled in emulsion droplets. *J. Mol. Catal. Chem.,* **2006**, *258*(1-2), 261-266.
        [http://dx.doi.org/10.1016/j.molcata.2006.05.058]

[76]   Huang, D.; Zhai, Z.; Lu, Y.C.; Yang, L.M.; Luo, G.S. Optimization of Composition of a Directly Combined Catalyst in Dibenzothiophene Oxidation for Deep Desulfurization. *Ind. Eng. Chem. Res.,* **2007**, *46*(5), 1447-1451.
        [http://dx.doi.org/10.1021/ie0611857]

[77]   Al-Shahrani, F.; Xiao, T.; Llewellyn, S.A.; Barri, S.; Jiang, Z.; Shi, H.; Martinie, G.; Green, M.L.H. Desulfurization of diesel *via* the H_2O_2 oxidation of aromatic sulfides to sulfones using a tungstate catalyst. *Appl. Catal. B,* **2007**, *73*(3-4), 311-316.
        [http://dx.doi.org/10.1016/j.apcatb.2006.12.016]

[78]   Te, M.; Fairbridge, C.; Ring, Z. Oxidation reactivities of dibenzothiophenes in polyoxometalate/H_2O_2 and formic acid/H_2O_2 systems. *Appl. Catal. A Gen.,* **2001**, *219*(1-2), 267-280.
        [http://dx.doi.org/10.1016/S0926-860X(01)00699-8]

[79]   Tam, P.S.; Kittrell, J.R.; Eldridge, J.W. Desulfurization of fuel oil by oxidation and extraction. 1. Enhancement of extraction oil yield. *Ind. Eng. Chem. Res.,* **1990**, *29*(3), 321-324.
        [http://dx.doi.org/10.1021/ie00099a002]

[80]   Zaykina, R.F.; Zaykin, Y.A.; Yagudin, S.G.; Fahruddinov, I.M. Radiat. Specific approaches to radiation processing of high-sulfuric oil. *Phys. Chem,* **2004**, *71*, 467-470.
        [http://dx.doi.org/10.1016/j.radphyschem.2004.04.077]

[81]   Ishihara, A.; Wang, D.; Dumeignil, F.; Amano, H.; Qian, E.W.; Kabe, T. Oxidative desulfurization and denitrogenation of a light gas oil using an oxidation/adsorption continuous flow process. *Appl. Catal. A Gen.,* **2005**, *279*(1-2), 279-287.
        [http://dx.doi.org/10.1016/j.radphyschem.2004.04.077]

[82]   Chan, N.Y.; Lin, T.Y.; Yen, T.F. Superoxides: Alternative Oxidants for the Oxidative Desulfurization Process. *Energy Fuels,* **2008**, *22*(5), 3326-3328.
[http://dx.doi.org/10.1021/ef800460g]

[83]   Bhutto, A.W.; Abro, R.; Gao, S.; Abbas, T.; Chen, X.; Yu, G. Oxidative desulfurization of fuel oils using ionic liquids: A review. *J. Taiwan Inst. Chem. Eng.,* **2016**, *62*, 84-97.
[http://dx.doi.org/10.1016/j.jtice.2016.01.014]

[84]   Houda, S.; Lancelot, C.; Blanchard, P.; Poinel, L.; Lamonier, C. Oxidative desulfurization of heavy oils with high sulfur content: a review. *Catalysts,* **2018**.
[http://dx.doi.org/10.3390/catal8090344]

[85]   Xie, L.L.; Favre-Reguillon, A.; Wang, X.X.; Fu, X.; Pellet-Rostaing, S.; Toussaint, G.; Geantet, C.; Vrinat, M.; Lemaire, M. Selective extraction of neutral nitrogen compounds found in diesel feed by 1-butyl-3-methyl-imidazolium chloride. *Green Chem.,* **2008**, *10*(5), 524-531.
[http://dx.doi.org/10.1039/b800789f]

[86]   Xie, L.L.; Favre-Reguillon, A.; Pellet-Rostaing, S.; Wang, X-X.; Fu, X.; Estager, J.; Vrinat, M.; Lemaire, M. Selective Extraction and Identification of Neutral Nitrogen Compounds Contained in Straight-Run Diesel Feed Using Chloride Based Ionic Liquid. *Ind. Eng. Chem. Res.,* **2008**, *47*(22), 8801-8807.
[http://dx.doi.org/10.1021/ie701704q]

[87]   Huh, E.S.; Zazybin, A.; Palgunadi, J.; Ahn, S.; Hong, J.; Kim, H.S.; Cheong, M.; Ahn, B.S. Zn-Containing Ionic Liquids for the Extractive Denitrogenation of a Model Oil: A Mechanistic Consideration. *Energy Fuels,* **2009**, *23*(6), 3032-3038.
[http://dx.doi.org/10.1021/ef900073a]

[88]   Toledo-Antonio, J.A.; Angeles-Chavez, C.; Cortes-Jacome, M.A.; Alvarez-Ramirez, F.; Ruiz-Morales, Y.; Ferrat-Torres, G.; Flores-Ortiz, L.F.; López Salinas, E.; Lozada y Cassou, M. Nanostructured titanium oxide material and its synthesis procedure. US Pat. 7,645,439. **2006**.

[89]   Laredo, G.C. Efecto de los compuestos nitrogenados característicos del diesel en la velocidad de hidrodesulfuración del dibenzotiofeno. *Universidad Autónoma Metropolitana, Mexico, 2001,* **2001**, 40-67.

[90]   Cerón, M.A.; Guzmán-Lucero, D.; Palomeque, J.F.; Martínez-Palou, R. Parallel Microwave-assisted Synthesis and Screening of Ionic Liquids for Denitrogenation of Straight-Run Diesel Feed by liquid-liquid extraction. *Comb. Chem. High Throughput Screen.,* **2012**, *15*, 427-432.
[http://dx.doi.org/10.2174/138620712800194477] [PMID: 22263864]

[91]   Laredo, G.C.; Likhanova, N.V.; Lijanova, I.V.; Rodriguez-Heredia, B.; Castillo, J.J.; Perez-Romo, P. Synthesis of ionic liquids and their use for extracting nitrogen compounds from gas oil feeds towards diesel fuel production. *Fuel Process. Technol.,* **2015**, *130*, 38-45.
[http://dx.doi.org/10.1016/j.fuproc.2014.08.025]

[92]   Laredo, G.C.; Vega-Merino, P.M.; Trejo-Zárraga, F.; Castillo, J. Denitrogenation of middle distillates using adsorbent materials towards ULSD production: A review. *Fuel Process. Technol.,* **2013**, *106*, 21-32.
[http://dx.doi.org/10.1016/j.fuproc.2012.09.057]

[93]   Hansmeier, A.R.; Meindersma, G.W.; de Haan, A.B. Desulfurization and denitrogenation of gasoline and diesel fuels by means of ionic liquids. *Green Chem.,* **2011**, *13*(7), 1907.
[http://dx.doi.org/10.1039/c1gc15196g]

[94]   Asumana, C.; Yu, G.; Guan, Y.; Yang, S.; Zhou, S.; Chen, X. Extractive denitrogenation of fuel oils with dicyanamide-based ionic liquids. *Green Chem.,* **2011**, *13*(11), 3300-3305.
[http://dx.doi.org/10.1039/c1gc15747g]

[95]   Fan, Y.; Cai, D.; Zhang, S.; Wang, H.; Guo, K.; Zhang, L.; Yang, L. Effective removal of nitrogen compounds from model diesel fuel by easy-to-prepare ionic liquids. *Separ. Purif. Tech.,* **2019**, *222*,

92-98.
[http://dx.doi.org/10.1016/j.seppur.2019.04.026]

[96]   Li, W.S.; Liu, J. Removal of basic nitrogen compounds from fuel oil with [C$_4$ mim]Br/ZnCl$_2$ ionic liquid. *Petrol. Sci. Technol.,* **2017**, *35*(13), 1364-1369.
[http://dx.doi.org/10.1080/10916466.2017.1331241]

[97]   Salleh, M.Z.M.; Hadj-Kali, M.K.; Hizaddin, H.F.; Ali Hashim, M. Extraction of nitrogen compounds from model fuel using 1-ethyl-3-methylimidazolium methanesulfonate. *Separ. Purif. Tech.,* **2018**, *196*, 61-70.
[http://dx.doi.org/10.1016/j.seppur.2017.07.068]

[98]   Zhou, Z.; Li, W.; Liu, J. Removal of basic nitrogen compounds from fuel oil with [Hnmp]H$_2$PO$_4$ ionic liquid. *Chem. Biochem. Eng. Q.,* **2017**, *31*(1), 63-68.
[http://dx.doi.org/10.15255/CABEQ.2016.955]

[99]   Verdía, P.; González, E.J.; Moreno, D.; Palomar, J.; Tojo, E. Deepening of the Role of Cation Substituents on the Extractive Ability of Pyridinium Ionic Liquids of N-Compounds from Fuels. *ACS Sustain. Chem.& Eng.,* **2017**, *5*(2), 2015-2025.
[http://dx.doi.org/10.1021/acssuschemeng.6b02922]

[100]   Zhou, Z.Q.; Li, W.S.; Liu, J. Removal of nitrogen compounds from fuel oils using imidazolium-based ionic liquids. *Petrol. Sci. Technol.,* **2017**, *35*(1), 45-50.
[http://dx.doi.org/10.1080/10916466.2016.1248771]

[101]   Vilas, M.; González, E.J.; Tojo, E. Extractive denitrogenation of model oils with tetraalkyl substituted pyridinium based ionic liquids. *Fluid Phase Equilib.,* **2015**, *396*, 66-73.
[http://dx.doi.org/10.1016/j.fluid.2015.03.032]

[102]   Abro, R.; Abro, M.; Gao, S.; Bhutto, A.W.; Ali, Z.M.; Shah, A.; Chen, X.; Yu, G. Extractive denitrogenation of fuel oils using ionic liquids: a review. *RSC Advances,* **2016**, *6*(96), 93932-93946.
[http://dx.doi.org/10.1039/C6RA09370A]

[103]   Prado, G.H.C.; Rao, Y.; de Klerk, A. Nitrogen removal from oil: A review. *Energy Fuels,* **2017**, *31*(1), 14-36.
[http://dx.doi.org/10.1021/acs.energyfuels.6b02779]

[104]   Hommeltoft, S.I. Isobutane alkylation. *Appl. Catal. A Gen.,* **2001**, *221*(1-2), 421-428.
[http://dx.doi.org/10.1016/S0926-860X(01)00817-1]

[105]   Miranda A.D.; Martínez-Palou,R.; Domínguez, J. M. Procedimiento para la remoción de contaminantes fluorados de efluentes de hidrocarburos. Alma D. Miranda, Rafael Martínez-Palou, J. M. Domínguez. IMP-1090. Folio. MX/a/2018/001281, Exp. MX/E/2018/007094. Patent pending (In Spanish).

[106]   Guzmán, A.; Zuazo, I.; Feller, A.; Olindo, R.; Sievers, C.; Lercher, J.A. Influence of the activation temperature on the physicochemical properties and catalytic activity of La-X zeolites for isobutane/cis-2-butene alkylation. *Microporous Mesoporous Mater.,* **2006**, *97*(1-3), 49-57.
[http://dx.doi.org/10.1016/j.micromeso.2006.08.006]

[107]   Feller, A.; Lercher, J.A. Chemistry and technology of isobutane/alkene alkylation catalyzed by liquid and solid acids. *Adv. Catal.,* **2004**, *48*, 229-295.
[http://dx.doi.org/10.1016/S0360-0564(04)48003-1]

[108]   Feller, A.; Guzmán, A.; Zuazo, I.; Lercher, J.A. On the mechanism of catalyzed isobutane/butene alkylation by zeolites. *J. Catal.,* **2004**, *224*(1), 80-93.
[http://dx.doi.org/10.1016/j.jcat.2004.02.019]

[109]   Feller, A.; Barth, J-O.; Guzmán, A.; Zuazo, I.; Lercher, J.A. Deactivation pathways in zeolite-catalyzed isobutane/butene alkylation. *J. Catal.,* **2003**, *220*(1), 192-206.
[http://dx.doi.org/10.1016/S0021-9517(03)00251-3]

[110]   Zhao, Z.; Sun, W.; Yang, X.; Ye, X.; Wu, Y. Study of the catalytic behaviors of concentrated

heteropolyacid solution. I. A novel catalyst for isobutane alkylation with butenes. *Catal. Lett.,* **2000,** *65*(1/3), 115-121.
[http://dx.doi.org/10.1023/A:1019009119808]

[111] Kumar, P.; Vermeiren, W.; Dath, J.P.; Hoelderich, W.F. Alkylation of raffinate II and isobutane on nafion silica nanocomposite for the production of isooctane. *Energy Fuels,* **2006,** *20*(2), 481-487.
[http://dx.doi.org/10.1021/ef050264c]

[112] Liu, Z.; Xu, C.; Huang, C. Method for manufacturing alkylate oil with composite ionic liquid used as catalyst. U.S. Pat. 0133056. **2004.**

[113] Wang, H.; Meng, X.; Zhao, G.; Zhang, S. Isobutane/butene alkylation catalyzed by ionic liquids: a more sustainable process for clean oil production. *Green Chem.,* **2017,** *19*(6), 1462-1489.
[http://dx.doi.org/10.1039/C6GC02791A]

[114] Salah, H.B.; Nancarrow, P.; Al-Othman, A. Ionic liquid-assisted refinery processes – A review and industrial perspective. *Fuel,* **2021,** *302*, 121195.
[http://dx.doi.org/10.1016/j.fuel.2021.121195]

CHAPTER 3

Application of Ionic Liquids in CO_2 Capture

Abstract: The oil industry is the industry that generates the most carbon dioxide (CO_2) worldwide, therefore the development of alternatives for the capture, use and transformation of CO_2 into products of greater added value is of great interest. This chapter presents an overview of ionic liquids application for CO_2 capture.

Keywords: Ionic liquids, CO_2, Environmental pollution, Absorption, Adsorption, Epoxides, Carbamates.

INTRODUCTION

The CO_2 emissions from the use of fossil fuels are a constant worry worldwide, for their concentration increase in the atmosphere is strictly related to the growing world energy demand and at the same time, it is one of the greenhouse effect gases with a higher impact on global warming and climatic change [1]. Although the development of new energy sources with low emissions of contaminants has to be a long-term goal in our societies, in the near future, the development of efficient technologies for the capture, storage and recycling of CO_2 is probably the only available strategy for controlling the CO_2 levels emitted into the atmosphere. On the other hand, oil refining is one of the industries that produce the highest CO_2 emissions and for this reason, the IMP is very interested in the research works on this topic.

The use of diverse technologies to capture, store and transform CO_2 could provide a mid-term solution to mitigate the environmental impact and allow the human society to keep on using fossil fuels as energy sources at least until renewable energy technologies become ready for their large-scale application [2].

From the Industrial Revolution, the atmospheric concentration of CO_2 has increased by approximately 35% of its original concentration, reaching a level above 400 ppm; for this reason, it has been considered one of the main climatic change factors and then, it is imperative that the atmospheric emissions be reduced, being one of the topics to which researchers have paid more attention in the current century.

The oil industry is one of the main sectors contributing to the generation of CO_2 and its capture and separation from other gases such as hydrogen and methane are vital. Until now, the development of efficient, economical and environmentally friendly technologies for separating these gases continues being one of the most important challenges faced by this industry.

The CO_2 capture technologies are commonly classified according to the process point where they are applied. Any technological planning involving the application of any of these alternatives will involve an increase in the energy demand of the process, given the energy required for CO_2 separation. Fig. (**3.1**) shows the schematic representation of different alternatives: pre-combustion, post-combustion and oxy-combustion [3].

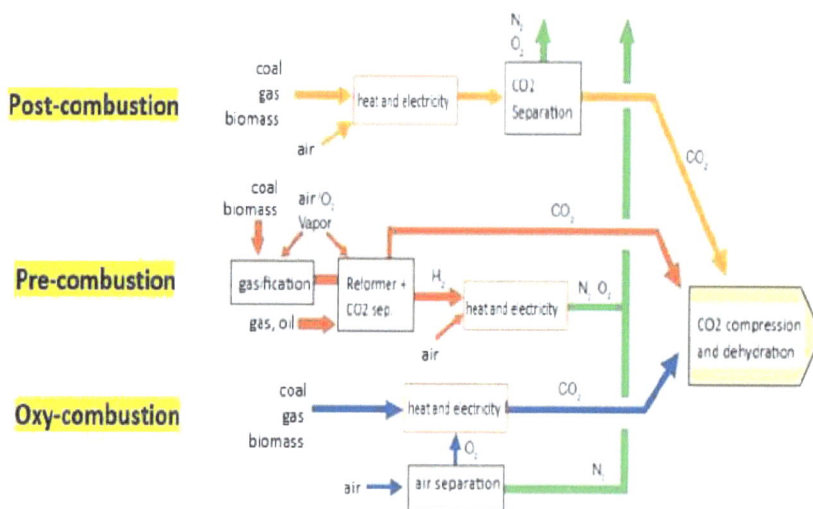

Fig. (3.1). Processes and CO_2 capture at different stages.

The post-combustion CO_2 capture technologies separate CO_2 from the gas current at the outlet of the conventional combustion process of fossil fuel. These technologies are located downstream in the process and keep on with the traditional installation of gas depuration equipment for a given pollutant, CO_2 in this case. Most industrial processes that require fossil fuels use air as a carburant; then, this technology can be applied directly without provoking alterations of the very process, which offers high flexibility and adaptation capacity to the installation operative conditions [4].

These systems start from a current with very low CO_2 concentration and a huge flow of combustion gases. Such characteristics in these systems require a large energy amount, which becomes evident in the high investment and operation costs of the currently developed technologies [5]. Fig. (**3.2**) shows the different available alternatives for the application of the post-combustion CO_2 capture technology [6].

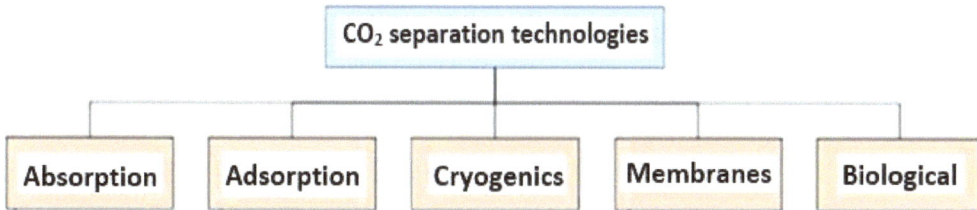

Fig. (**3.2**). Main CO_2 capture technologies.

In general, the methods to remove CO_2 from gaseous effluents have to comply with the following characteristics: high absorption capacity, fast kinetics, capability of being regenerated and stable through time within a wide operation range [7].

Despite all the efforts, there is not a scaled CO_2 capture procedure yet that could be economically viable, complying with all the previously described requisites; for this reason, many government programs with important investment have been promoted to attend this problem that continues to be a very active research topic.

CO_2 Capture with ILs

The CO_2 capture mechanisms depend on the chemical structure of the employed materials and their properties. In this context, ILs are among the most promising chemical products. These liquids can ease the CO_2 capture without losing solvent in the gas current. ILs have low toxicity and volatility and can be designed with the right properties for efficient CO_2 capture. The evident interest in using ILs for this purpose is confirmed by the increasing number of scientific reports in this area (Fig. **3.3**).

In the last years, multiple studies have been carried out to explore the perspectives of ILs to be applied in the separation of gases. In the case of ILs for capturing CO_2, the search has been focused on not-supported ILs [8 - 19], and on ILs supported on membranes (SILMs) [20 - 30].

Year	Value
2019	57
2018	82
2017	84
2016	100
2015	79
2014	106
2013	76
2012	68
2011	44
2010	35

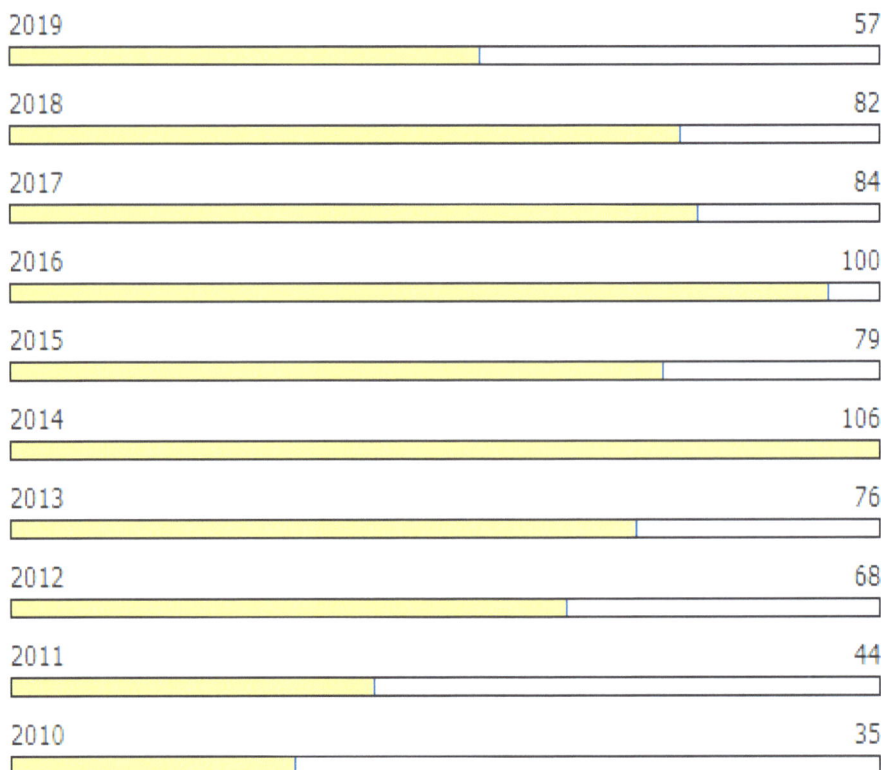

Fig. (3.3). Papers published in the last years on CO_2 capture employing ILs. (Source: SciFinder).

Another interesting alternative for CO_2 capture is represented by poly(ionic liquid)s (polyILs). ILs with the right structure can be polymerized through the cation and/or anion, forming solid films. PolyILs combine the exceptional properties of ILs with the flexibility and properties of the macromolecular architecture and provide new properties and functions that have huge potential in many applications [31 - 38].

Functionalized ILs for CO_2 Capture

In the last years, a new strategy based on chemical absorption has been described to increase the CO_2 capture yield of ILs as a function of the nature of the cationic and anionic groups.

The most known strategy has been the introduction of an amino group into the cation, anion or both, which allows the CO_2 capture through the formation of a covalent bond between the gas carbon atom and amino group(s) present in the IL (Fig. **3.4**) [39 - 42].

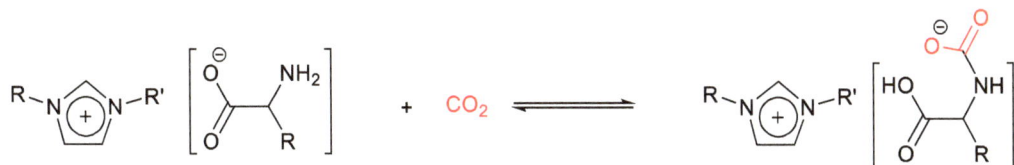

Fig. (3.4). CO_2 capture by chemisorption through the use of ILs containing amino acids in the anion.

Amino-functionalized ILs, especially those containing amino groups derived from amino acids (AAs), have displayed excellent capacity for capturing CO_2 through chemical absorption [43 - 49]. Recently, the IMP published the article Absorption of CO_2 with amino acid-based ionic liquids and the corresponding amino acid precursors. J. Guzmán-Pantoja, C. Ortega-Guevara, R. García de León, R. Martínez-Palou. Chem. Eng. Technol. 2017, 40, 2339-2345 [50]. In this work, a comparative study between ILs derived from AAs and the very precursor AAs as CO_2 absorbents was carried out. The study revealed that AAs displayed excellent performance in the CO_2 capture, and for this reason, their direct use can be a good alternative, especially when the AAs are dissolved in water and the solution pH is adjusted to its basic isoelectric point with ammonia.

The ILs were synthesized from commercial precursors such as tetramethylammonium hydroxide (**1**) and tetrabutylphosphonium hydroxide (**2**) by ionic exchange with the corresponding AAs (**3**) as shown in Fig. (**3.5**).

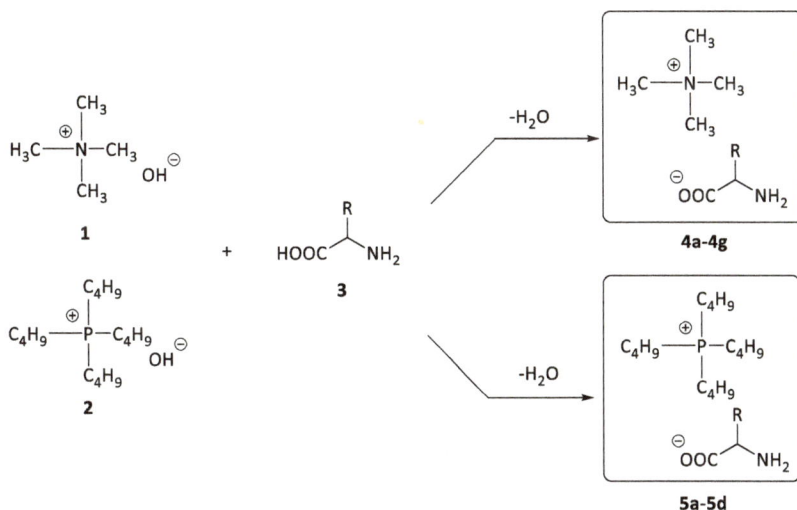

Fig. (3.5). Reaction scheme for the synthesis of ILs with anions derived from AAs and the cations tetramethylammonium (**4a-4g**) and tetrabutylphosphonium (**5a-5d**).

Table **3.1** shows the performance of the ILs synthesized in the mentioned work in comparison with methylamine (MEA, entry 12), which has been a widely used absorbent in the oil industry.

Table 3.1. CO_2 absorption at 303 K using ILs containing the cations tetramethylammonium (TMA) and tetrabutylphosphonium (TBP) and AAs as anions.

Entry	IL	CO_2 Absorption (mmol CO_2 mol^{-1} IL)		
		Pure IL	IL in aqueous solution (30%)	IL in aqueous solution (3%)
1	[TMA][Gly]	334.8	416.3	886.8
2	[TMA][Ala]	460.1	531.6	1293.8
3	[TMA][Val]	405.4	303.3	2048.6
4	[TMA][Glu]	219.9	306.4	2429.3
5	[TMA][Hys]	445.4	515.5	2175.4
6	[TMA][Arg]	549.2	898.4	1917.5
7	[TMA][Lys]	613.6	811.5	2089.0
8	[TBP][Gly]	946.6	1143.1	2733.2
9	[TBP][Ala]	1032.3	1185.4	3232.7
10	[TBP][Arg]	1523.1	2156.3	4791.3
11	[TBP][Lys]	1594.6	2267.7	5019.1
12	MEA	--	470.0	--

As it can be seen in Table **3.1**, the IL [TBP][Lys] (entry 11) was the absorbent that showed the best performance as CO_2 absorbent, both pure and in an aqueous solution. On the other hand, the precursor AAs of these ILs was evaluated at the same temperature. In order to have a higher absorption capacity in the case of the AA solution, it is important that the amine group be in its neutral from; for this purpose, a base has to be added, and in this case, NaOH was used to establish a pH equal to the amine pK. In this case, lysine resulted to be the AA with the best capture capacity (1902.8 mmol of CO_2 mol^{-1} per mol of AA in aqueous solution at 3%).

Finally, in this study, a microwave-assisted method for the regeneration of ILs in the saturated CO_2 solution was described for the first time. With this procedure, the AA/IL solutions can be reused for 10 absorption-desorption cycles without evident original absorption capacity loss.

One of the main problems that limit the yield capacity of the ILs is that they have high viscosity, and this one is increased, in general, after absorbing a certain amount of CO_2; for this reason, one of the main strategies has been the use of ILs with either solvent or amine mixtures.

In order to face the high viscosity of the ILs with amino functionality during the CO_2 capture, another strategy employing an amino-functionalized-IL mixture blended with other absorbents such as SiO_2, monoethanolamine, diethanolamine, polyethylene glycol and DBU, among others, has been studied [51 - 55].

Another aspect that has limited the industrial application of ILs is their cost, which in conventional ILs oscillates around 1-10 USD/g, about 100 times more expensive than common organic solvents; in addition, they are not commercially available on a high scale. This cost is increased in the case of functionalized ILs that require more complex synthesis processes. Up to now, only protic ILs obtained from amines through neutralization reactions, capable of being totally regenerated after many cycles could be competitive for industrial-scale applications aimed at capturing CO_2 [56, 57].

Separation and Capture of CO_2 by Means of Supported-ILs Membranes

In general, supported liquid membranes consist of two phases: in general, a porous support and a liquid solvent occluded in the pores. In supported liquid membranes, solvent loss can occur due to volatilization when conventional solvents are employed. However, when ILs are used as the liquid phase, the resulting membranes (SILMs) are very stable due to the low vapor pressure and high chemical and thermal stability displayed by ILs. There are other SILM variants that have been conceived to improve even more the stability and useful life of the membranes like composite IL/polymer membranes, poly(IL) membranes and composite gel/IL membranes. These membranes have not only been studied for carrying out the selective separation of CO_2 from other gases, but also to perform the separation of volatile organic compounds (VOCs) [58].

SILMs increase the efficiency and selectivity of separation processes with respect to non-functionalized liquid membranes due to their high surface area per volume unit, which favors better contact between the ILs and phases to be separated. SILMs combine extraction and desorption processes, which requires less solvent amount than conventional extraction processes. On the other hand, SILMs can keep their performance for many hours even with temperature and/or pressure increments and in general, they possess considerable thermal stability.

The different ILs immobilized on supported membranes can be classified according to their interaction with the supporting material. This interaction can be of two types: 1) by physical adsorption, which is a simple impregnation without chemical bonds between the IL and solid support and is achieved by direct immersion either with pressure or vacuum [59], and 2) by chemical adsorption, where the IL functional groups form covalent bonds with the structure of the porous support [60].

The studies on SILMs have been mainly focused on the CO_2 capture from methane or nitrogen mixtures [61], but these membranes have also been proposed to separate other gases from hydrocarbon blends of great interest in the oil area such as sulfur dioxide, carbon monoxide, hydrogen, nitrogen, water (dehydration) and separation of olefins/paraffins.

Several research groups have proposed predictive correlations of the gas solubility in ILs [62 - 69], and gas diffusivity in ILs [70 - 75], based on their physicochemical properties. According to Robeson, the separation factor for the gas pairs varies inversely with the permeability of the most permeable gas of a given pair. In this equilibrium relationship, the limit superior can be represented in a lol-log plot of αij (separation factor = Pi/Pj) *versus* Pi, where Pi is the permeability of the most permeable gas. The slope of this line (N) can be related to the difference between the molecular diameters of the gases, where the molecular diameter is the Lennard-Jones kinetic diameter. The analysis of literature data was carried out to assess the progress of gas separation polymeric membranes with respect to the limit superior to the Robeson line.

In 2008, Robeson reviewed his theory on the limit superior using available data. The results show modest changes from 1991. Some perfluorinated polymer membranes [76] display data slightly above the limit superior for the CO_2/N_2 and CO_2/CH_4 mixtures.

Among the high variety of existing ILs, it has been shown that those based on imidazole salts are more effective in separating CO_2 and CH_4 and the CO_2/N_2 pair [77, 78].

Likewise, comparative experiments have been carried out on the influence of the cation on the separation of CO_2 from gas mixtures, finding once again that imidazole salts are the best option for this separation type, regarding both permeability and selectivity, following in efficiency the ILs from the ammonium cation and after those based on the cation phosphonium [79].

Also, studies have been conducted on the effect of the IL structure on the efficiency of such separation [80], being the main tested ILs those based on the 1-ethyl-3methylimidazole [EMIM] and 1-butyl-3-methylimidazole [BMIM] structures as cations, varying the anionic part with structures such as [TfO]⁻, [Tf₂N]⁻, [BF₄]⁻, and [PF₆]⁻, among others [81].

The IL membrane technology for the separation of different gas and liquid mixtures, including CO_2, was reviewed in 2014 in the compendium: Ionic Liquid Membranes for the Separation of Gases and Liquids: An overview. R. Martínez-Palou, N. V. Likhanova, O. Olivares-Xomelt. Pet. Chem. 2014, 54, 595-607 [82]. In this work, the great potential displayed by IL membranes separating CO_2 from other gases, with which they form common mixtures in the oil industry, was confirmed. The IL membranes can be of different types, and some of them are produced by anchoring the IL on a given inorganic material, displaying high efficiency (Fig. **3.6**).

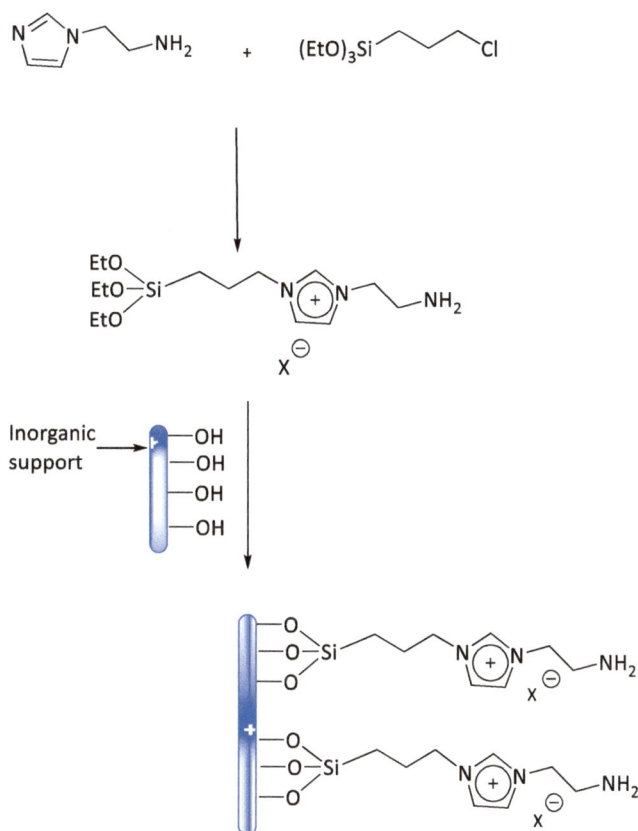

Fig. (3.6). Strategy example for producing membranes of ILs supported on inorganic materials.

By means of the article: CO_2/N_2 separation using alumina supported membranes based on new functionalized ionic liquids by C. E. Sánchez-Fuentes, D. Guzmán-Lucero, M. Torres-Rodríguez, N. V. Likhanova, J. Navarrete-Bolaños, O. Olivares-Xometl, I. V. Lijanova. Sep. Sci. Technol. 2017, 182, 59-68 [83], the IMP published a new method for the separation of CO_2 and N_2 employing IL functionalized membranes. In this work, compatible liquid membranes based on three ILs were prepared: (1-(2-aminoethyl)-3-methylimidazolium trifluorome-thanesulfonate ([AEMIm]Tf), methylimidazolium 1-(2-aminoethyl)-3- tetra-fluoroborate ([AEMIm]BF$_4$) and trioctylmethylammonium anthranilate ([TOMA]An), which were used as mobile phases. The permeation tests were performed with pure gases at 30 °C and 1 bar. The IL functionalized membranes having amino groups in the cation ([AEMIm]BF$_4$ and [AEMIm]Tf), and [TOMA]An as anion reacted with CO$_2$ to produce carbamates. [TOMA]An, which was synthesized by ionic exchange with the anthranilic acid from the IL previously prepared from trioctylamine and dimethylcarbonate in methanol (Fig. **3.7**), displayed the best permselectivity, reaching selectivity equal to 70, thus surpassing Robeson's barrier.

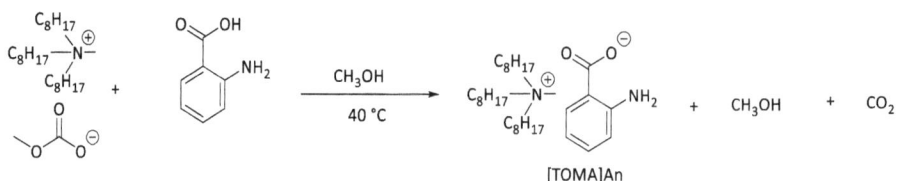

Fig. (3.7). Synthesis of [TOMA]An.

Currently, the IMP is preparing the publication of the work XXX, where the permeability and selectivity of SILMs containing anions consisting of AAs synthesized in our laboratories, which were supported on polyimide semi-permeable membranes, were studied. These membranes showed selectivity toward the separation of CO$_2$ mixed with other gases, such as hydrogen and methane, from gaseous effluents produced under post-combustion operation conditions that are typical in the oil industry. The membrane containing the IL 1-butyl-3-methylimidazolium as cation with valinate ([BMIM][Val]) as anion dissolved in glycerol (50% w/w) displayed the best selectivity (1.47 at 1 atm).

Transformation and Valorization of CO$_2$

Up to now, the capture and storage of CO$_2$ are the most studied alternatives to solve the problem concerning the emissions of this greenhouse effect gas; notwithstanding, the valorization of CO$_2$ is an interesting option from the economic standpoint and currently, various industrial processes are benefited by the high availability of this gas [84 - 86]. The *in situ* transformation of CO$_2$ into chemical products or commercial interest materials represents a promising approach that has been taken into account by researchers working on this topic over the last decades [87 - 89].

One of the problems that have limited the valorization of CO_2 is that this compound has a very stable chemical structure, for it is the most oxidized carbon structure, transforming it into a molecule with high thermal stability; this characteristic prevents its reactivity and as a consequence, many evaluated valorization processes have given relatively low yields of the desired products and high energy demand reaction conditions and very expensive catalysts are required [90].

For this reason, the reactions involving highly reactive substrates such as epoxides and aziridines have been widely studied, for CO_2 reacts well with these strong nucleophiles to form new C-C and C-H bonds.

One of the alternatives that have drawn more attention in the transformation of CO_2 into added-value compounds is the production of carbonates, which are important raw materials for the synthesis of polycarbonates and polyurethanes, and of components of lithium ion batteries and aprotic polar solvents, among other applications [91]. Cyclic carbonates can be synthesized by cycloaddition of CO_2 to oxiranes, using catalysts that include metal complexes from the main group, simple alkaline metal salts, ammonium salts, phosphines and oxidative and not oxidative transition metal complexes [92].

Also, the IMP has explored the valorization of CO_2 by transforming it into carbonates. In addition to the previously discussed work by Likhanova *et al.* in 2007, in the work Efficient synthesis of Organic Carbonates to the multigram level from CO_2 using a new chitin-supported catalyst by H. Díaz-Velázquez; J. Guzmán-Pantoja, R. García de León, R. Martínez-Palou. Catal. Lett. 2017, 147, 2260-2268 [93], a methodology for producing organic carbonates by reacting CO_2 with different epoxides, employing potassium iodide supported on chitin as a catalyst with a DBU base, is described.

In addition to be a very efficient alternative for transforming CO_2 into low toxicity products, organic carbonates are important monomers for producing commercially valuable polymers such as polycarbonates and polyurethanes; and also, they are employed as lithium battery components [94].

Table **3.2** shows the results regarding the effect exerted by different catalysts that were evaluated in the transformation of cyclohexene oxide into its corresponding epoxide employing CO_2 as a reactant; it can be observed that the KI/DBU/chitin mixture was the most effective catalyst for this reaction.

Table 3.2. Effect of different catalysts evaluated in the cyclohexene oxide reaction.

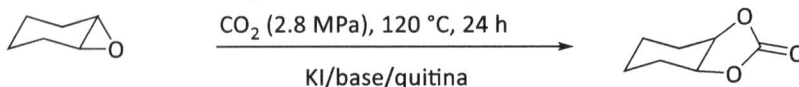

Entry	Catalyst	Yield (%)	pKa[a]
1	KI[b]	0.5	-
2	DBU[c]	61	13.5
3	KI/chitin[d]	46	-
4	chitin/DBU	75	13.5
5	KI/chitin/DBU	82	13.5
6	KI/chitin/triethylamine	47	10.7
7	KI/chitin/1-methylimidazole	75	7.1
8	KI/chitin/diethanolamine	44	8.9
9	KI/chitin/dimethylaminopyridine	70	9.6
10	KI/chitin/aminoguanidine bicarbonate	46	11.1
11	KI/chitin/1,1,3,3-tetramethylguanidine	72	13

[a]pKa of the conjugated acid in water.

CONCLUDING REMARKS

As described in this chapter, ILs have great potential for CO_2 capture. Supported or polymerized amino-functionalized ILs show higher values in carbon dioxide absorption or separation processes. Those containing amino acids, most of all lysine, in the anion structure have demonstrated excellent capture properties and recoverability, while IL with anthranilate anion displayed the permselectivity equal to 70 in the CO_2/N_2 mixture.

Further studies are required to bring this technology to fruition on a large scale, mostly by forming value-added products like cyclic carbonates, which is expected to happen in the near future.

REFERENCES

[1]   Koytsoumpa, E.I.; Bergins, C.; Kakaras, E. The CO_2 economy: Review of CO_2 capture and reuse technologies. *J. Supercrit. Fluids,* **2018**, *132*, 3-16.
[http://dx.doi.org/10.1016/j.supflu.2017.07.029]

[2]   Rafiee, A.; Rajab Khalilpour, K.; Milani, D.; Panahi, M. Trends in CO_2 conversion and utilization: A review from process systems perspective. *J. Environ. Chem. Eng.,* **2018**, *6*(5), 5771-5794.
[http://dx.doi.org/10.1016/j.jece.2018.08.065]

[3]   Figueroa, J.D.; Fout, T.; Plasynski, S.; McIlvried, H.; Srivastava, R.D. Advances in CO_2 capture technology—The U.S. Department of Energy's Carbon Sequestration Program. *Int. J. Greenh. Gas*

Control, **2008**, *2*(1), 9-20.
[http://dx.doi.org/10.1016/S1750-5836(07)00094-1]

[4]   MacDowell, N.; Florin, N.; Buchard, A.; Hallett, J.; Galindo, A.; Jackson, G.; Adjiman, C.S.; Williams, C.K.; Shah, N.; Fennell, P. An overview of CO2 capture technologies. *Energy Environ. Sci.,* **2010**, *3*(11), 1645-1669.
[http://dx.doi.org/10.1039/c004106h]

[5]   Feron, P.H.M.; Hendriks, C.A. CO $_2$ Capture Process Principles and Costs. *Oil Gas Sci. Technol.,* **2005**, *60*(3), 451-459.
[http://dx.doi.org/10.2516/ogst:2005027]

[6]   Wang, M.; Lawal, A.; Stephenson, P.; Sidders, J.; Ramshaw, C. Post-combustion CO₂ capture with chemical absorption: A state-of-the-art review. *Chem. Eng. Res. Des.,* **2011**, *89*(9), 1609-1624.
[http://dx.doi.org/10.1016/j.cherd.2010.11.005]

[7]   Althuluth, M.; Overbeek, J.P.; van Wees, H.J.; Zubeir, L.F.; Haije, W.G.; Berrouk, A.; Peters, C.J.; Kroon, M.C. Natural gas purification using supported ionic liquid membrane. *J. Membr. Sci.,* **2015**, *484*, 80-86.
[http://dx.doi.org/10.1016/j.memsci.2015.02.033]

[8]   Lei, Z.; Han, J.; Zhang, B.; Li, Q.; Zhu, J.; Chen, B. Solubility of CO₂ in binary mixture of room-temperature ionic liquids at high pressures. *J. Chem. Eng. Data,* **2012**, *57*(8), 2153-2159.
[http://dx.doi.org/10.1021/je300016q]

[9]   Zeng, S.; Zhang, X.; Bai, L.; Zhang, X.; Wang, H.; Wang, J.; Bao, D.; Li, M.; Liu, X.; Zhang, S. Ionic-Liquid-Based CO $_2$ Capture Systems: Structure, Interaction and Process. *Chem. Rev.,* **2017**, *117*(14), 9625-9673.
[http://dx.doi.org/10.1021/acs.chemrev.7b00072] [PMID: 28686434]

[10]   Baltus, R.E.; Counce, R.M.; Culbertson, B.H.; Luo, H.; DePaoli, D.W.; Dai, S.; Duckworth, D.C. Examination of the Potential of Ionic Liquids for Gas Separations. *Sep. Sci. Technol.,* **2005**, *40*(1-3), 525-541.
[http://dx.doi.org/10.1081/SS-200042513]

[11]   Shiflett, M.B.; Yokozeki, A. Solubilities and Diffusivities of Carbon Dioxide in Ionic Liquids: [bmim][PF $_6$] and [bmim][BF $_4$]. *Ind. Eng. Chem. Res.,* **2005**, *44*(12), 4453-4464. [bmim]. [BF4].
[http://dx.doi.org/10.1021/ie058003d]

[12]   Ebner, A.D.; Ritter, J.A. State-of-the-art Adsorption and Membrane Separation Processes for Carbon Dioxide Production from Carbon Dioxide Emitting Industries. *Sep. Sci. Technol.,* **2009**, *44*(6), 1273-1421.
[http://dx.doi.org/10.1080/01496390902733314]

[13]   Shokouhi, M.; Adibi, M.; Jalili, A.H.; Hosseini-Jenab, M.; Mehdizadeh, A. Solubility and Diffusion of H $_2$ S and CO $_2$ in the Ionic Liquid 1-(2-Hydroxyethyl)-3-methylimidazolium Tetrafluoroborate. *J. Chem. Eng. Data,* **2010**, *55*(4), 1663-1668.
[http://dx.doi.org/10.1021/je900716q]

[14]   Ramdin, M.; de Loos, T.W.; Vlugt, T.J.H. State-of-the-Art of CO $_2$ Capture with Ionic Liquids. *Ind. Eng. Chem. Res.,* **2012**, *51*(24), 8149-8177.
[http://dx.doi.org/10.1021/ie3003705]

[15]   Aki, S.N.V.K.; Mellein, B.R.; Saurer, E.M.; Brennecke, J.F. High-pressure phase behavior of carbon dioxide with imidazolium-based ionic liquids. *J. Phys. Chem. B,* **2004**, *108*(52), 20355-20365.
[http://dx.doi.org/10.1021/jp046895+]

[16]   Anthony, J.L.; Anderson, J.L.; Maginn, E.J.; Brennecke, J.F. Anion effects on gas solubility in ionic liquids. *J. Phys. Chem. B,* **2005**, *109*(13), 6366-6374.
[http://dx.doi.org/10.1021/jp046404l] [PMID: 16851709]

[17]   Wappel, D.; Gronald, G.; Kalb, R.; Draxler, J. Ionic liquids for post-combustion CO2 absorption. *Int.*

J. Greenh. Gas Control, **2010**, *4*(3), 486-494.
[http://dx.doi.org/10.1016/j.ijggc.2009.11.012]

[18]   Soutullo, M.D.; Odom, C.I.; Wicker, B.F.; Henderson, C.N.; Stenson, A.C.; Davis, J.H. Reversible CO$_2$ Capture by Unexpected Plastic-, Resin-, and Gel-like Ionic Soft Materials Discovered during the Combi-Click Generation of a TSIL Library. *Chem. Mater.,* **2007**, *19*(15), 3581-3583.
[http://dx.doi.org/10.1021/cm0705690]

[19]   Wang, C.; Luo, X.; Zhu, X.; Cui, G.; Jiang, D.; Deng, D.; Li, H.; Dai, S. The strategies for improving carbon dioxide chemisorption by functionalized ionic liquids. *RSC Advances,* **2013**, *3*(36), 15518-155270.
[http://dx.doi.org/10.1039/c3ra42366b]

[20]   Zhang, Y.; Zhang, S.; Lu, X.; Zhou, Q.; Fan, W.; Zhang, X. Dual amino-functionalised phosphonium ionic liquids for CO$_2$ capture. *Chemistry,* **2009**, *15*(12), 3003-3011.
[http://dx.doi.org/10.1002/chem.200801184] [PMID: 19185037]

[21]   Zhang, J.; Zhang, S.; Dong, K.; Zhang, Y.; Shen, Y.; Lv, X. Supported absorption of CO$_2$ by tetrabutylphosphonium amino acid ionic liquids. *Chemistry,* **2006**, *12*(15), 4021-4026.
[http://dx.doi.org/10.1002/chem.200501015] [PMID: 16528787]

[22]   Li, B.; Duan, Y.; Luebke, D.; Morreale, B. Advances in CO$_2$ capture technology: A patent review. *Appl. Energy,* **2013**, *102*, 1439-1447.
[http://dx.doi.org/10.1016/j.apenergy.2012.09.009]

[23]   Swati, I.K.; Sohaib, Q.; Cao, S.; Younas, M.; Liu, D.; Gui, J.; Rezakazemi, M. Protic/aprotic ionic liquids for effective CO$_2$ separation using supported ionic liquid membrane. *Chemosphere,* **2021**, *267*, 128894.
[http://dx.doi.org/10.1016/j.chemosphere.2020.128894] [PMID: 33187654]

[24]   Malik, M.A.; Hashim, M.A.; Nabi, F. Ionic liquids in supported liquid membrane technology. *Chem. Eng. J.,* **2011**, *171*(1), 242-254.
[http://dx.doi.org/10.1016/j.cej.2011.03.041]

[25]   Lozano, L.J.; Godínez, C.; de los Ríos, A.P.; Hernández-Fernández, F.J.; Sánchez-Segado, S.; Alguacil, F.J. Recent advances in supported ionic liquid membrane technology. *J. Membr. Sci.,* **2011**, *376*(1-2), 1-14.
[http://dx.doi.org/10.1016/j.memsci.2011.03.036]

[26]   Hasib-ur-Rahman, M.; Siaj, M.; Larachi, F. Ionic liquids for CO2 capture—Development and progress. *Chem. Eng. Process.,* **2010**, *49*(4), 313-322.
[http://dx.doi.org/10.1016/j.cep.2010.03.008]

[27]   Bara, J.E.; Camper, D.E.; Gin, D.L.; Noble, R.D. Room-temperature ionic liquids and composite materials: platform technologies for CO$_2$) capture. *Acc. Chem. Res.,* **2010**, *43*(1), 152-159.
[http://dx.doi.org/10.1021/ar9001747] [PMID: 19795831]

[28]   Camper, D.; Bara, J.E.; Gin, D.L.; Noble, R.D. Room-temperature ionic liquid-amine solutions: tunable solvents for efficient and reversible capture of CO$_2$. *Ind. Eng. Chem. Res.,* **2008**, *47*(21), 8496-8498.
[http://dx.doi.org/10.1021/ie801002m]

[29]   Zhang, X.; Zhang, X.; Dong, H.; Zhao, Z.; Zhang, S.; Huang, Y. Carbon capture with ionic liquids: overview and progress. *Energy Environ. Sci.,* **2012**, *5*(5), 6668-6681.
[http://dx.doi.org/10.1039/c2ee21152a]

[30]   Rubin, E.S.; Mantripragada, H.; Marks, A.; Versteeg, P.; Kitchin, J. The outlook for improved carbon capture technology. *Pror. Energy Combust. Sci.,* **2012**, *38*(5), 630-671.
[http://dx.doi.org/10.1016/j.pecs.2012.03.003]

[31]   Jiayin, Y.; Markus, A. Poly(ionic liquid)s: Polymers expanding classical property profiles. *Polymers (Basel),* **2011**, *52*, 1469-1482.

[32] Bara, J.E.; Hatakeyama, E.S.; Gin, D.L.; Noble, R.D. Improving CO_2 permeability in polymerized room-temperature ionic liquid gas separation membranes through the formation of a solid composite with a room-temperature ionic liquid. *Polym. Adv. Technol.,* **2008**, *19*(10), 1415-1420.
[http://dx.doi.org/10.1002/pat.1209]

[33] Bara, J.E.; Gin, D.L.; Noble, R.D. Effect of anion on gas separation performance of polymer-roo--temperature ionic liquid composite membranes. *Ind. Eng. Chem. Res.,* **2008**, *47*(24), 9919-9924.
[http://dx.doi.org/10.1021/ie801019x]

[34] Bara, J.E.; Noble, R.D.; Gin, D.L. Effect of "Free" Cation Substituent on Gas Separation Performance of Polymer–Room-Temperature Ionic Liquid Composite Membranes. *Ind. Eng. Chem. Res.,* **2009**, *48*(9), 4607-4610.
[http://dx.doi.org/10.1021/ie801897r]

[35] Xie, Y.; Liang, J.; Fu, Y.; Huang, M.; Xu, X.; Wang, H.; Tu, S.; Li, J. Hypercrosslinked mesoporous poly(ionic liquid)s with high ionic density for efficient CO_2 capture and conversion into cyclic carbonates. *J. Mater. Chem. A Mater. Energy Sustain.,* **2018**, *6*(15), 6660-6666.
[http://dx.doi.org/10.1039/C8TA01346B]

[36] Zhou, X.; Weber, J.; Yuan, J. Poly(ionic liquid)s: Platform for CO_2 capture and catalysis. *Curr. Opin. Green Sustain. Chem.,* **2019**, *16*, 39-46.
[http://dx.doi.org/10.1016/j.cogsc.2018.11.014]

[37] Raja Shahrom, M.S.; Wilfred, C.D.; MacFarlane, D.R.; Vijayraghavan, R.; Chong, F.K. Amino acid based poly(ionic liquid) materials for CO2 capture: Effect of anion. *J. Mol. Liq.,* **2019**, *276*, 644-652.
[http://dx.doi.org/10.1016/j.molliq.2018.12.044]

[38] Carlisle, T.K.; Wiesenauer, E.F.; Nicodemus, G.D.; Gin, D.L.; Noble, R.D. Ideal CO_2/Light Gas Separation Performance of Poly(vinylimidazolium) Membranes and Poly(vinylimidazolium)-Ionic Liquid Composite Films. *Ind. Eng. Chem. Res.,* **2013**, *52*(3), 1023-1032.
[http://dx.doi.org/10.1021/ie202305m]

[39] Cevasco, G.; Chiappe, C. Are ionic liquids a proper solution to current environmental challenges? *Green Chem.,* **2014**, *16*(5), 2375-2385.
[http://dx.doi.org/10.1039/c3gc42096e]

[40] Solangi, N.H.; Anjum, A.; Tanjung, F.A.; Mazari, S.A.; Mubarak, N.M. A review of recent trends and emerging perspectives of ionic liquid membranes for CO_2 separation. *J. Environ. Chem. Eng.,* **2021**, *9*(5), 105860.
[http://dx.doi.org/10.1016/j.jece.2021.105860]

[41] Goodrich, B.F.; de la Fuente, J.C.; Gurkan, B.E.; Zadigian, D.J.; Price, E.A.; Huang, Y.; Brennecke, J.F. Experimental Measurements of Amine-Functionalized Anion-Tethered Ionic Liquids with Carbon Dioxide. *Ind. Eng. Chem. Res.,* **2011**, *50*(1), 111-118.
[http://dx.doi.org/10.1021/ie101688a]

[42] Shiflett, M.B.; Drew, D.W.; Cantini, R.A.; Yokozeki, A. Carbon Dioxide Capture Using Ionic Liquid 1-Butyl-3-methylimidazolium Acetate. *Energy Fuels,* **2010**, *24*(10), 5781-5789.
[http://dx.doi.org/10.1021/ef100868a]

[43] Soutullo, M.D.; Odom, C.I.; Wicker, B.F.; Henderson, C.N.; Stenson, A.C.; Davis, J.H. Reversible CO₂ Capture by Unexpected Plastic-, Resin-, and Gel-like Ionic Soft Materials Discovered during the Combi-Click Generation of a TSIL Library. *Chem. Mater.,* **2007**, *19*(15), 3581-3583.
[http://dx.doi.org/10.1021/cm0705690]

[44] Zhao, Y.; Dong, Y.; Guo, Y.; Huo, F.; Yan, F.; He, H. Recent progress of green sorbents-based technologies for low concentration CO_2 capture. *Chin. J. Chem. Eng.,* **2021**, *31*, 113-125.
[http://dx.doi.org/10.1016/j.cjche.2020.11.005]

[45] Gurkan, B.E.; de la Fuente, J.C.; Mindrup, E.M.; Ficke, L.E.; Goodrich, B.F.; Price, E.A.; Schneider, W.F.; Brennecke, J.F. Equimolar CO_2 absorption by anion-functionalized ionic liquids. *J. Am. Chem.*

Soc., **2010**, *132*(7), 2116-2117.
[http://dx.doi.org/10.1021/ja909305t] [PMID: 20121150]

[46]   Chen, J.J.; Li, W.W.; Li, X.L.; Yu, H.Q. Carbon dioxide capture by aminoalkyl imidazolium-based ionic liquid: a computational investigation. *Phys. Chem. Chem. Phys.,* **2012**, *14*(13), 4589-4596.
[http://dx.doi.org/10.1039/c2cp23642g] [PMID: 22358056]

[47]   Wang, C.; Luo, H.; Jiang, D.; Li, H.; Dai, S. Carbon dioxide capture by superbase-derived protic ionic liquids. *Angew. Chem. Int. Ed.,* **2010**, *49*(34), 5978-5981.
[http://dx.doi.org/10.1002/anie.201002641] [PMID: 20632428]

[48]   Wang, C.; Luo, X.; Luo, H.; Jiang, D.; Li, H.; Dai, S. Tuning the basicity of ionic liquids for equimolar CO_2 capture. *Angew. Chem. Int. Ed.,* **2011**, *50*(21), 4918-4922.
[http://dx.doi.org/10.1002/anie.201008151] [PMID: 21370373]

[49]   Feng, Z.; Cheng-Gang, F.; You-Ting, W.; Yuan-Tao, W.; Ai-Min, L.; Zhi-Bing, Z. Absorption of CO_2 in the aqueous solutions of functionalized ionic liquids and MDEA. *Chem. Eng. J.,* **2010**, *160*(2), 691-697.
[http://dx.doi.org/10.1016/j.cej.2010.04.013]

[50]   Guzmán, J.; Ortega-Guevara, C.; de León, R.G.; Martínez-Palou, R. Absorption of CO_2 with Amino Acid-Based Ionic Liquids and Corresponding Amino Acid Precursors. *Chem. Eng. Technol.,* **2017**, *40*(12), 2339-2345.
[http://dx.doi.org/10.1002/ceat.201600593]

[51]   Li, X.; Hou, M.; Zhang, Z.; Han, B.; Yang, G.; Wang, X.; Zou, L. Absorption of CO_2 by ionic liquid/polyethylene glycol mixture and the thermodynamic parameters. *Green Chem.,* **2008**, *10*(8), 879-884.
[http://dx.doi.org/10.1039/b801948g]

[52]   Liu, F.; Shen, Y.; Shen, L.; Sun, C.; Chen, L.; Wang, Q.; Li, S.; Li, W. Novel Amino-Functionalized Ionic Liquid/Organic Solvent with Low Viscosity for CO_2 Capture. *Environ. Sci. Technol.,* **2020**, *54*(6), 3520-3529.
[http://dx.doi.org/10.1021/acs.est.9b06717] [PMID: 32062963]

[53]   Zhu, X.; Chen, Z.; Ai, H. Amine-functionalized ionic liquids for CO_2 capture. *J. Mol. Model.,* **2020**, *26*(12), 345.
[http://dx.doi.org/10.1007/s00894-020-04563-6] [PMID: 33215296]

[54]   Wang, C.; Luo, H.; Luo, X.; Li, H.; Dai, S. Equimolar CO_2 capture by imidazolium-based ionic liquids and superbase systems. *Green Chem.,* **2010**, *12*(11), 2019-2023.
[http://dx.doi.org/10.1039/c0gc00070a]

[55]   Hanioka, S.; Maruyama, T.; Sotani, T.; Teramoto, M.; Matsuyama, H.; Nakashima, K.; Hanaki, M.; Kubota, F.; Goto, M. CO_2 separation facilitated by task-specific ionic liquids using a supported liquid membrane. *J. Membr. Sci.,* **2008**, *314*(1-2), 1-4.
[http://dx.doi.org/10.1016/j.memsci.2008.01.029]

[56]   Chen, Y.; Cao, Y.; Sun, X.; Yan, C.; Mu, T. New criteria combined of efficiency, greenness, and economy for screening ionic liquids for CO2 capture. *Int. J. Greenh. Gas Control,* **2013**, *16*, 13-20.
[http://dx.doi.org/10.1016/j.ijggc.2013.02.025]

[57]   Aghaie, M.; Rezaei, N.; Zendehboudi, S. A systematic review on CO_2 capture with ionic liquids: Current status and future prospects. *Renew. Sustain. Energy Rev.,* **2018**, *96*, 502-525.
[http://dx.doi.org/10.1016/j.rser.2018.07.004]

[58]   Yahaya, G.O.; Hamad, F.; Bahamdan, A.; Tammana, V.V.R.; Hamad, E.Z. Supported ionic liquid membrane and liquid–liquid extraction using membrane for removal of sulfur compounds from diesel/crude oil. *Fuel Process. Technol.,* **2013**, *113*, 123-129.
[http://dx.doi.org/10.1016/j.fuproc.2013.03.028]

[59]   Hernández-Fernández, F.J.; de los Ríos, A.P.; Tomás-Alonso, F.; Palacios, J.M.; Víllora, G.

Preparation of supported ionic liquid membranes: Influence of the ionic liquid immobilization method on their operational stability. *J. Membr. Sci.,* **2009**, *341*(1-2), 172-177.
[http://dx.doi.org/10.1016/j.memsci.2009.06.003]

[60]   Valkenberg, M.H.; deCastro, C.; Hölderich, W.F. Immobilisation of ionic liquids on solid supports. *Green Chem.,* **2002**, *4*(2), 88-93.
[http://dx.doi.org/10.1039/b107946h]

[61]   Zheng, S.; Zeng, S.; Li, Y.; Bai, L.; Bai, Y.; Zhang, X.; Liang, X.; Zhang, S. State of the art of ionic liquid-modified adsorbents for CO$_2$ capture and separation. *AIChE J.,* **2022**, *682*, e17500.

[62]   Song, Z.; Shi, H.; Zhang, X.; Zhou, T. Prediction of CO2 solubility in ionic liquids using machine learning methods. *Chem. Eng. Sci.,* **2020**, *223*, 115752.
[http://dx.doi.org/10.1016/j.ces.2020.115752]

[63]   Camper, D.; Becker, C.; Koval, C.; Noble, R. Low pressure hydrocarbon solubility in room temperature ionic liquids containing imidazolium rings interpreted using regular solution theory. *Ind. Eng. Chem. Res.,* **2005**, *44*(6), 1928-1933.
[http://dx.doi.org/10.1021/ie049312r]

[64]   Kilaru, P.K.; Condemarin, R.A.; Scovazzo, P. Correlations of low-pressure carbon dioxide and alkenes solubilities in imidazolium-, phosphonium-, and ammonium-based room temperature ionic liquids. Part I: using surface tension. *Ind. Eng. Chem. Res.,* **2008**, *47*(3), 900-909.
[http://dx.doi.org/10.1021/ie070834r]

[65]   Kilaru, P.K.; Scovazzo, P. Correlations of low-pressure carbon dioxide and alkenes solubilities in imidazolium-, phosphonium-, and ammonium-based room temperature ionic liquids. Part II. Using activation energy of viscosity. *Ind. Eng. Chem. Res.,* **2008**, *47*(3), 910-919.
[http://dx.doi.org/10.1021/ie070836b]

[66]   Zhang, X.; Liu, Z.; Wang, W. Screening of ionic liquids to capture CO$_2$ by COSMO-RS and experiments. *AIChE J.,* **2008**, *54*(10), 2717-2728.
[http://dx.doi.org/10.1002/aic.11573]

[67]   Carlisle, T.K.; Bara, J.E.; Gabriel, C.J.; Noble, R.D.; Gin, D.L. Interpretation of CO$_2$ solubility and selectivity in nitrile-functionalized room-temperature ionic liquids using a group contribution approach. *Ind. Eng. Chem. Res.,* **2008**, *47*(18), 7005-7012.
[http://dx.doi.org/10.1021/ie8001217]

[68]   Schilderman, A.M.; Raeissi, S.; Peters, C.J. Solubility of carbon dioxide in the ionic liquid 1-ethyl-3-methylimidazolium bis(trifluoromethylsulfonyl)imide. *Fluid Phase Equilib.,* **2007**, *260*(1), 19-22.
[http://dx.doi.org/10.1016/j.fluid.2007.06.003]

[69]   Sprunger, L.M.; Proctor, A.; Acree, W.E., Jr; Abraham, M.H. LFER correlations for room temperature ionic liquids: Separation of equation coefficients into individual cation-specific and anion-specific contributions. *Fluid Phase Equilib.,* **2008**, *265*(1-2), 104-111.
[http://dx.doi.org/10.1016/j.fluid.2008.01.006]

[70]   Morgan, D.; Ferguson, L.; Scovazzo, P. Diffusivity of gases in room temperature ionic liquids: data and correlation obtained using a lag-time technique. *Ind. Eng. Chem. Res.,* **2005**, *44*(13), 4815-4823.
[http://dx.doi.org/10.1021/ie048825v]

[71]   Camper, D.; Becker, C.; Koval, C.; Noble, R. Diffusion and solubility measurements in room temperature ionic liquids. *Ind. Eng. Chem. Res.,* **2006**, *45*(1), 445-450.
[http://dx.doi.org/10.1021/ie0506668]

[72]   Hou, Y.; Baltus, R.E. Experimental measurement of the solubility and diffusivity of CO$_2$ in room-temperature ionic liquids using a transient thin-liquid-film method. *Ind. Eng. Chem. Res.,* **2007**, *46*(24), 8166-8175.
[http://dx.doi.org/10.1021/ie070501u]

[73]   Ferguson, L.; Scovazzo, P. Solubility, diffusivity, and permeability of gases in phosphonium-based

room temperature ionic liquids: data and correlations. *Ind. Eng. Chem. Res.,* **2007**, *46*(4), 1369-1374.
[http://dx.doi.org/10.1021/ie0610905]

[74]   Condemarin, R.; Scovazzo, P. Gas permeabilities, solubilities, diffusivities, and diffusivity correlations for ammonium-based room temperature ionic liquids with comparison to imidazolium and phosphonium RTIL data. *Chem. Eng. J.,* **2009**, *147*(1), 51-57.
[http://dx.doi.org/10.1016/j.cej.2008.11.015]

[75]   Robeson, L.M. Correlation of separation factor *versus* permeability for polymeric membranes. *J. Membr. Sci.,* **1991**, *62*(2), 165-185.
[http://dx.doi.org/10.1016/0376-7388(91)80060-J]

[76]   Robeson, L.M. The upper bound revisited. *J. Membr. Sci.,* **2008**, *320*(1-2), 390-400.
[http://dx.doi.org/10.1016/j.memsci.2008.04.030]

[77]   Scovazzo, P.; Havard, D.; McShea, M.; Mixon, S.; Morgan, D. Long-term, continuous mixed-gas dry fed CO_2/CH_4 and CO_2/N_2 separation performance and selectivities for room temperature ionic liquid membranes. *J. Membr. Sci.,* **2009**, *327*(1-2), 41-48.
[http://dx.doi.org/10.1016/j.memsci.2008.10.056]

[78]   Uchytil, P.; Schauer, J.; Petrychkovych, R.; Setnickova, K.; Suen, S.Y. Ionic liquid membranes for carbon dioxide–methane separation. *J. Membr. Sci.,* **2011**, *383*(1-2), 262-271.
[http://dx.doi.org/10.1016/j.memsci.2011.08.061]

[79]   Hayashi, E.; Hashimoto, K.; Thomas, L. M.; Tsuzuki, S.; Watanabe, M. Role of Cation Structure in CO_2 Separation by Ionic Liquid/Sulfonated Polyimide Composite Membrane. *Membranes,* **2019**, *9*, 81.
[http://dx.doi.org/10.3390/membranes9070081]

[80]   Cichowska-Kopczyńska, I.; Joskowska, M.; Dębski, B.; Łuczak, J.; Aranowski, R. Influence of Ionic Liquid Structure on Supported Ionic Liquid Membranes Effectiveness in Carbon Dioxide/Methane Separation. *J. Chem.,* **2013**, *2013*, 1-10.
[http://dx.doi.org/10.1155/2013/980689]

[81]   Iarikov, D.D.; Hacarlioglu, P.; Oyama, S.T. Supported room temperature ionic liquid membranes for CO_2/CH_4 separation. *Chem. Eng. J.,* **2011**, *166*(1), 401-406.
[http://dx.doi.org/10.1016/j.cej.2010.10.060]

[82]   Martínez-Palou, R.; Likhanova, N.V.; Olivares-Xometl, O. Supported ionic liquid membranes for separations of gases and liquids: an overview. *Petrol. Chem.,* **2014**, *54*(8), 595-607.
[http://dx.doi.org/10.1134/S0965544114080106]

[83]   Sánchez-Fuentes, C.E.; Guzmán-Lucero, D.; Torres-Rodríguez, M.; Likhanova, N.V.; Navarrete-Bolaños, J.; Olivares-Xometl, O.; Lijanova, I.V. CO_2/N_2 separation using alumina supported membranes based on new functionalized ionic liquids. *Sep. Sci. Technol.,* **2017**, *182*, 59-68.

[84]   Liu, Q.; Wu, L.; Jackstell, R.; Beller, M. Using carbon dioxide as a building block in organic synthesis. *Nat. Commun.,* **2015**, *6*(1), 5933.
[http://dx.doi.org/10.1038/ncomms6933] [PMID: 25600683]

[85]   Sakakura, T.; Choi, J.C.; Yasuda, H. Transformation of carbon dioxide. *Chem. Rev.,* **2007**, *107*(6), 2365-2387.
[http://dx.doi.org/10.1021/cr068357u] [PMID: 17564481]

[86]   Aresta, M.; Dibenedetto, A. Utilisation of CO_2 as a chemical feedstock: opportunities and challenges. *Dalton Trans.,* **2007**, *28*(28), 2975-2992.
[http://dx.doi.org/10.1039/b700658f] [PMID: 17622414]

[87]   Aresta, M.; Dibenedetto, A.; Tommasi, I. Developing Innovative Synthetic Technologies of Industrial Relevance Based on Carbon Dioxide as Raw Material. *Energy Fuels,* **2001**, *15*(2), 269-273.
[http://dx.doi.org/10.1021/ef000242k]

[88]   Aresta, M.; Dibenedetto, A.; Angelini, A. Catalysis for the valorization of exhaust carbon: from CO_2 to chemicals, materials, and fuels. technological use of CO_2. *Chem. Rev.,* **2014**, *114*(3), 1709-1742.

[http://dx.doi.org/10.1021/cr4002758] [PMID: 24313306]

[89]   Mikkelsen, M.; Jørgensen, M.; Krebs, F.C. The teraton challenge. A review of fixation and transformation of carbon dioxide. *Energy Environ. Sci.,* **2010**, *3*(1), 43-81.
[http://dx.doi.org/10.1039/B912904A]

[90]   Aresta, M.; Dibenedetto, A.; Quaranta, E. State of the art and perspectives in catalytic processes for CO_2 conversion into chemicals and fuels: The distinctive contribution of chemical catalysis and biotechnology. *J. Catal.,* **2016**, *343*, 2-45.
[http://dx.doi.org/10.1016/j.jcat.2016.04.003]

[91]   North, M.; Pasquale, R.; Young, C. Synthesis of cyclic carbonates from epoxides and CO_2. *Green Chem.,* **2010**, *12*(9), 1514-1539.
[http://dx.doi.org/10.1039/c0gc00065e]

[92]   Artz, J.; Müller, T.E.; Thenert, K.; Kleinekorte, J.; Meys, R.; Sternberg, A.; Bardow, A.; Leitner, W. Sustainable Conversion of Carbon Dioxide: An Integrated Review of Catalysis and Life Cycle Assessment. *Chem. Rev.,* **2018**, *118*(2), 434-504.
[http://dx.doi.org/10.1021/acs.chemrev.7b00435] [PMID: 29220170]

[93]   Díaz Velázquez, H.; Guzmán Pantoja, J.; Meneses Ruiz, E.; García de León, R.; Martínez Palou, R. Efficient synthesis of Organic Carbonates to the multigram level from CO_2 using a new chitin-supported catalyst. *Catal. Lett.,* **2017**, *147*(9), 2260-2268.
[http://dx.doi.org/10.1007/s10562-017-2123-4]

[94]   Omae, I. Aspects of carbon dioxide utilization. *Catal. Today,* **2006**, *115*(1-4), 33-52.
[http://dx.doi.org/10.1016/j.cattod.2006.02.024]

CHAPTER 4

Application of ILs in the Breaking of Emulsions Found in the Oil Industry

Abstract: Emulsions are commonly found in oil and cause major operational problems, so emulsion breakage is a major issue in this industry. Among the alternatives for breaking emulsions is the use of de-emulsifying products and among them, some ILs with amphiphilic properties have gained an important place by demonstrating a very good efficiency as emulsion breakers in combination with other methods such as heating.

Keywords: Amphiphilic, Demulsifiers, Emulsions, Ionic liquids, Microwave, O/W, W/O.

INTRODUCTION

Emulsions are of great importance in the oil industry, for naturally, oil is found as an emulsion. Due to the presence of congenital water in oil wells, natural crude oil emulsifiers and temperature and pressure conditions, the formation of simple and complex emulsions is favored; the most common emulsion is the one where oil is the continuous phase with emulsified water drops within. The emulsion is the result of the coproduction of water in the oil reservoir.

An emulsion is a lyophobic colloid (a solution that cannot be formed by spontaneous dispersion) and has a dimension close to 1000 nm. When a system consisting of surfactant, water and oil is stirred, one of the liquid phases is dispersed as drops (with diameters from 1 to 100 μm), thus producing an emulsion [1]. The aim of employing surfactants is, on the one hand, to ease the extension of the interface during the emulsion formation process, and on the other hand, to stabilize the emulsion by retarding the coalescence of the dispersed phase drops.

Emulsions are stabilized by the presence of an emulsifying agent, which is normally an amphipathic species that forms a surface film on the interface between each colloidal drop and the dispersion medium, thus reducing the interfacial tension and preventing coagulation.

Rafael Martínez Palou & Natalya V. Likhanova

There are three essential requisites for an emulsion to be formed [2];

- Two immiscible liquids like water and oil.
- Enough stirring to disperse one of the liquids as droplets within the other liquid.
- An emulsifying agent for stabilizing the drops dispersed within the continuous phase.

In the oil industry and specialized literature, English nomenclature is used to define emulsions. The two immiscible liquids are referred to as water (W) and oil (O), either these are the very liquids or they represent the polar and nonpolar phases.

The most common emulsions are those with water drops dispersed within the oil phase and are referred to as water in oil (water / oil) emulsions and less common are those known as inverse emulsions, where oil is the dispersed phase and are referred to as oil in water (oil/water) emulsions (Fig. **4.1**).

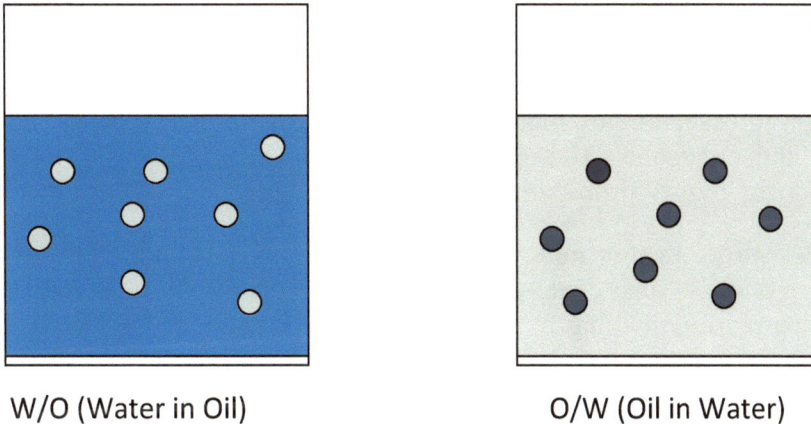

W/O (Water in Oil)   O/W (Oil in Water)

Fig. (4.1). Schematic representation of the two types of emulsions found in the oil industry.

W/O (Water in Oil) O/W (Oil in Water)

There are also complex emulsions, *e.g.* the dispersion of oil drops in water drops, which in turn are dispersed in a continuous oil phase (oil/water/oil; O/ W/O). The type of the formed emulsion depends on different factors.

Three main criteria are necessary for the formation of a crude oil emulsion: the contact of two immiscible liquids, a surfactant component working as an emulsifier agent and enough stirring for dispersing a liquid within the other as drops.

Some emulsions are separated in their water and oil phases once removed from the sea surface whereas the most stable emulsions can last for days and even years. Stability is a consequence of the small drop size and the presence of an interfacial film of emulsion drops that make stable dispersions [3].

On the other hand, colloidal species can be gathered in different ways and three processes are considered to reach stability:

1. Creaming: This phenomenon is the opposite to sedimentation and is the result of the formation of a kind of cream with different density between the two liquid phases, which creates a concentration gradient of drops, two or more drops are grouped, being in contact in just certain points and practically without a change on the total surface. Creaming is generally considered undesirable, as it makes storing and manipulation difficult, but it can be useful in special cases, mainly when an emulsion is to be concentrated. As it has been already stated, the migration process is referred to as creaming as long as the substance particles remain separated. This is what establishes the ideal difference between flocculation (where the particles are grouped) and emulsion breaking (where the particles get together).

2. Aggregation: It occurs when drops are very close for a long time without attraction forces acting on them. The species preserve their identity, but lose their kinetic independence. The segregation of droplets can lead to fusion and formation of bigger drops until the phase is separated.

3. Coalescence: Two or more drops get together to form a bigger unit with total surface reduction. The coalescence mechanism occurs through two stages: film drainage and film breaking.

These and other aspects related to the emulsion theory in the oil industry are discussed in Chapter 11 of the book 'Ionic liquids as Surfactants. Applications as demulsifiers of petroleum emulsions by Rafael Martínez-Palou and Jorge Aburto, in the book: Ionic Liquids, Current State of the Art. Intech. 2015, p. 305-326 [4].

Different studies have demonstrated that the physicochemical action mechanism of dehydrating or demulsifying agents is associated with the optimal formulation of the system (SAD = 0, where SAD stands for Surfactant Affinity Difference) [5].

The optimal formulation is defined basically as an equilibrium state between the surfactant affinities to aqueous and oil phases. The effects exerted by the different formulation variables (salinity, ACN, EON, WOR and temperature, among others) on the hydrophilic/lipophilic balance between the surfactant and its physicochemical environment have been established quantitatively [6].

In a surfactant-water-oil system, the optimal formulation is achieved when, through a unidimensional scan of any formulation variable, the system displays either a minimal or ultra-low interfacial tension, accompanied, in general, by the presence of a three-phase system, where most part of the surfactant is found in the intermediate phase. In the case of water in oil emulsions, it is not common to observe such a three-phase system and instability is detected by the coalescence progress and evolution of the dynamic interfacial tension [7].

Parameters that Play a Role in the Demulsification Process

There are different parameters that can reduce or increase the stability of an emulsion; some of these parameters are as follows:

Salinity

The increase in the water salinity stabilizes a W/O emulsion, making its breaking difficult. This salinity increase also provokes a hydrophilia reduction in the surfactants [8].

Temperature

A temperature increase reduces the viscosity of the oil phase, diminishing the difference between the densities of the phases, and favors the increase in the drop collisions, thus weakening the stability of the film surrounding the drops; therefore, the temperature increase tends to reduce the stability of the W/O emulsions [9].

pH

The hydrophilia of the surfactants tends to be increased when the emulsion pH is also increased. W/O emulsions can be produced in acid media (low pH), whereas O/W emulsions are more commonly developed in basic media (high pH). On the other hand, the asphaltenic interfacial films are more rigid in an acid medium and are gradually weakened as pH is increased; the opposite occurs with resin interfaces, but in general, acid pH values destabilize the W/O emulsions, easing their breaking [10].

Particle Size

Both droplet size and distribution influence the emulsion viscosity; smaller droplet size and narrow drop size distribution of a generic emulsion result in an emulsion with higher viscosity and thus more stability. This is obvious since it takes longer for the smaller drops to coalesce and eventually sediment (water globules) or float (oil droplets). An efficient demulsifier facilitates the coalescence of dispersed drops [11].

Water Content

The increase in the water content of an emulsion in the presence of demulsifiers promotes the emulsion breaking by diminishing its stability. As a consequence, the increase in water content reduces the emulsion breaking time and augments its viscosity [12].

Stirring Rate

A higher stirring rate or longer mixing time of O/W emulsions leads to the reduction of the average size distribution of dispersed oil droplets and an increase of both emulsion viscosity and stability due to improved particle-particle interactions as a consequence of higher interfacial area. More specifically, in systems containing small dispersed droplets (diameter < 1 lm), the colloidal surface forces and Brownian motion forces prevail over the hydrodynamic forces [13].

Other factors such as crude oil composition, specially the resin and asphaltene contents, which are natural emulsifiers that stabilize emulsions and considerably increase the crude oil viscosity [14, 15], solid contents [16], and emulsion age also affect significantly the emulsion stability [17].

Commercial Demulsifiers for Breaking W/O Emulsions

In general, commercial demulsifiers are mixtures of various components with different chemical structures and polymeric materials with a wide distribution of molecular weights. They consist of the active compound (from 30 to 50%) and the addition of suitable solvents such as aromatic naphtha and alcohols [18].

Copolymers in ethylene oxide and propylene oxide blocks, alkyl-pheno-
-formaldehyde resins, polyamines, fatty alcohols, oxyalkyl amines and polyester amines and their mixtures are among the most common compounds used as demulsifiers. These surfactants exert three main effects once they are adsorbed on the water-oil interface: one is the inhibition of the formation of a rigid film, another is the weakening of the film, making it compressible and the most important one is the system formulation change to reach the condition of SAD = 0, where SAD stands for Surfactant Affinity Difference [19].

Fig. (**4.2**) shows some structures of the commercial chemical compounds employed in the formulations of commercial demulsifiers. Usually, these compounds have one or several of these active ingredients in a suitable solvent. In the case of polymers, their molecular weight and chain distribution play a major role in the demulsifying effect.

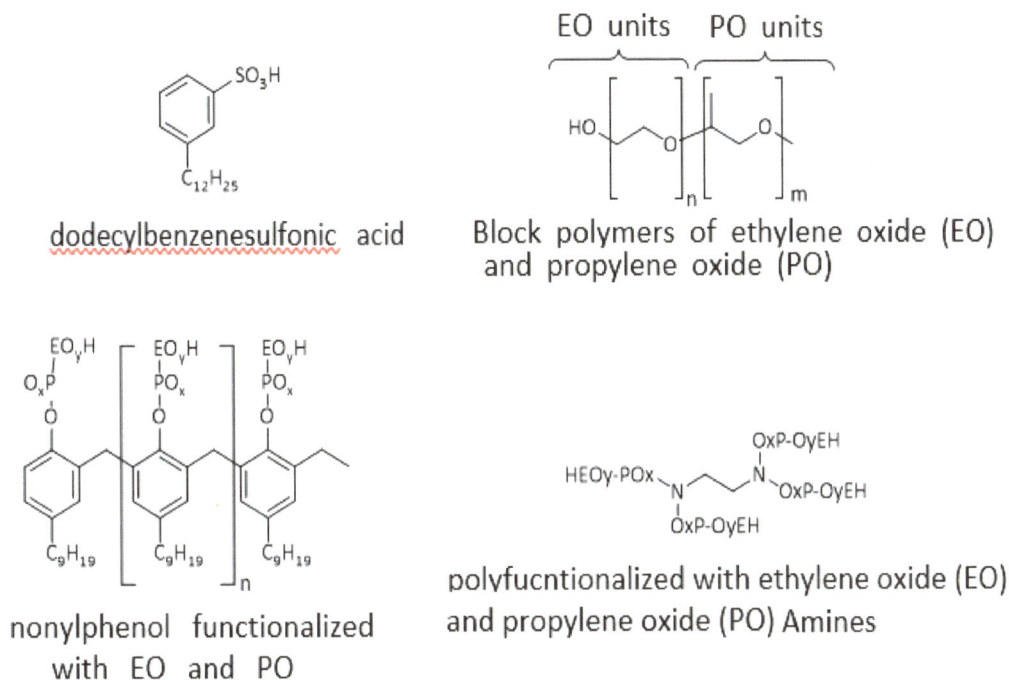

Fig. (4.2). Chemical structure of some typical commercial demulsifiers.

EO units, PO units; Block polymers of ethylene oxide (EO) and propylene oxide (PO); dodecylbenzenesulfonic acid; nonylphenol functionalized with EO and PO; Amines polyfucntionalized with ethylene oxide (EO) and propylene oxide (PO).

ILs as Demulsifying Agents of W/O Emulsions

ILs have been tested successfully as breaking agents of W/O emulsions. The most studied ILs for this application have been those containing the imidazolium cation with a long alkyl chain, which provides amphiphilic properties to this type of ILs [20 - 24].

In this context, the IMP has developed some prototypes with excellent performance as demulsifying agents of water-in-oil emulsions, which are also known as direct emulsions, especially when employing ILs.

In the article by Guzmán-Lucero D, Flores P, Rojo T, and Martínez-Palou, R, entitled Evaluation of ionic liquids as demulsifiers of water-in-crude oil emulsions. Study of microwave effect (Energy Fuel 2010; 24:3610-3615) [25], ILs based on ammonium, imidazole and pyridinium were synthesized and the potential of these compounds to break direct emulsions (W/O) was shown for the first time. To this end, three Mexican crude oils with different API gravities were evaluated: an intermediate one (API = 29.59°), a heavy one (API = 21.27°) and an extra-heavy one (API = 9.88°). It was observed that all the ILs demulsified suitably the intermediate oil; some of them broke correctly the heavy-crude-oil emulsions, but only trioctylmethylammonium chloride displayed efficient performance dehydrating the extra-heavy crude oil according to the bottle tests carried out at 80°C and employing the demulsifiers at 1000 ppm (Table **4.1**).

Table 4.1. Evaluation of the demulsifier performance of different ILs (1000 ppm).[a, b].

Demulsifier Code	Structure	Bottle Test Results		
		Intermediate Crude Oil	Heavy Crude Oil	Extra-heavy Crude Oil
TEA-C6-Br	$C_2H_5-\overset{\overset{C_2H_5}{\oplus}}{\underset{C_2H_5}{N}}-C_6H_{13}$ $\ominus$ Br	☺	☻	☻
TEA-C12-Br	$C_2H_5-\overset{\overset{C_2H_5}{\oplus}}{\underset{C_2H_5}{N}}-C_{12}H_{25}$ $\ominus$ Br	☺	☺ / ☻	☻
TEA-C18-Br	$C_2H_5-\overset{\overset{C_2H_5}{\oplus}}{\underset{C_2H_5}{N}}-C_{18}H_{37}$ $\ominus$ Br	☺	☻	☻
MIM-C14-Br	imidazolium $-N \oplus N-C_{14}H_{29}$ $\ominus$ Br	☺	☺ / ☻	☻
MIM-C18-Br	imidazolium $-N \oplus N-C_{18}H_{37}$ $\ominus$ Br	☺	☻	☻
Py-C14 Br	pyridinium $\oplus - C_{14}H_{29}$ $\ominus$ Br	☺	☺ / ☻	☻

(Table 4.1) cont.....

Demulsifier Code	Structure	Bottle Test Results		
		Intermediate Crude Oil	Heavy Crude Oil	Extra-heavy Crude Oil
Py-C18 Br		☺	☻	☻
TPACl		☺	☺	☻
THACl		☺	☺	☺ / ☻
TOACl		☺	☺	☺

[a] ☺: Good demulsification and well-defined separation of phases. ☺ / ☻: Partial separation of phases with slight interface definition. ☻: No separation of phases was observed. [b] Bottle test at 80°C for 10 h.

The use of microwave irradiation accelerated and significantly increased the demulsification efficiency of the ultra-heavy crude oil emulsion (Table **4.2**).

Table **4.2**. Microwave (MW) demulsification of ultra-heavy crude oil using TOACl at different concentrations.

Demulsifier (Concentration)	Temperature (°C)	Water Removal by MW Irradiation (%)		
		10 min	20 min	30 min
--	80	0	0	0
TOACl (1000 ppm)	80	15	88	89
TOACl (1500 ppm)	80	16	90	95
TOACl (1500 ppm)	100	21	94	98
TOACl (2500 ppm)	100	20	90	95

Some authors have employed only the microwave energy (without adding chemical products) as an alternative to facilitate the breaking of emulsions [26 - 36], and in turn, other authors have conjugated the use of ILs as demulsifiers and microwaves as a heating source to reduce time and water separation efficiency as previously commented. As stated in the opinion article Applications of microwave for breaking petroleum emulsions by Martínez-Palou, R. Curr. Microwave Chem. 2017, 4, 276 [37], the synergistic effect exerted by ILs and microwaves on the water-oil demulsification process can be understood by describing two phenomena that take place simultaneously:

1. A considerable viscosity reduction occurs as the sample temperature is increased; in this sense, it is worth noting that microwaves produce the dielectric heating of the sample due to the friction among molecules. This heating process occurs much faster than conventional heating, for it does not depend on the thermal conduction properties of the recipient containing the sample. In contrast, conventional heating takes place by the conduction/convection mechanism, *i.e.* the heating source, first, has to transmit the heat to the container and this one to the sample; on the other hand, dielectric heating happens by direct interaction between molecules and radiation; for this reason, the heating process is accelerated and more homogeneous. Another significant difference is that dielectric heating is transmitted from the interior of the sample out so that the highest temperatures are located in the most internal zones of the irradiated sample. The presence of ILs creates an environment with higher polarity for hydrocarbons (low polarity medium) and then, the interaction between microwaves and medium is more efficient.

2. Microwaves induce the molecular rotation with which the emulsion Zeta potential is neutralized. The Zeta potential is the electric potential that exists in the particle shear plane at a short distance from the surface. The colloidal particles dispersed in a solution are electrically charged and the development of an electric charge network on the particle surface can affect the distribution of ions in a neighboring interfacial region and provoke an increase in the concentration of ions with charge opposite to that of particles close to the surface. Since microwaves induce the molecular rotation, they affect the order of the charges that generate the electrical double layer with which the Zeta potential is neutralized, favoring the emulsion collapse [38].

Binner *et al.* studied the water separation mechanism using microwaves, coming to the conclusion that the improved phase separation rates can be rationalized as a function of the unique thermal gradients that occur with microwave heating and their further impact on the viscosity and interfacial tension in the water/oil interface; in no case, evidences of non-thermal microwave effects were observed [39].

In the article Amphiphilic Choline Carboxylate Ionic Liquids as Demulsifiers of Water-in-Crude Oil Emulsions by the IMP researchers J. Aburto, D. M. Márquez, J. C. Navarro, and R. Martínez-Palou. Tenside Surf. Deterg. 2014, 51, 314-317 [40], once again, the efficiency of some ammonium-based ILs as demulsifying agents was shown by employing choline carboxylates.

In this work, four amphiphilic choline carboxylates were synthesized by ionic exchange from choline chloride (vitamin B4) and fatty acid salts under microwave irradiation according to the procedure described in Fig. (**4.3**).

Choline chloride

R = $C_{11}H_{23}$
R = $C_{13}H_{27}$
R = $C_{15}H_{31}$
R = $C_{17}H_{36}$

MW, 2.5 minutes

Choline carboxilate

Fig. (4.3). Microwave synthesis of choline carboxylates.

These anionic surfactants are derived from natural products, and for this reason, they can be considered as environmentally friendly. The evaluation of these products breaking water in crude oil emulsions using short intervals of microwave dielectric heating allowed the tracking down of the demulsification kinetics and the evaluation results were validated and confirmed by the classical procedure known as "bottle test". Choline palmitate displayed the best demulsifying performance of the Mexican heavy-crude-oil emulsion.

Ammonium salts also prompted dehydrating studies of crude oil by another IMP research team formed by Flores, C. A., Flores, E. A., Hernández, E., Castro, L. V., García, A., Alvarez, F., and Vázquez, F. S. The results were published in the paper titled Anion and cation effects of ionic liquids and ammonium salts evaluated as dehydrating agents for super-heavy crude oil: Experimental and theoretical points of view. J. Mol. Liq. 2014, 196, 249–257 [41], where the effect of the cation and anion in ammonium-based ILs belonging to two families, TOA (**1a-1c**) and OCD (**2a-2c**) (Fig. **4.4**), on the performance as dehydrating agents of two Mexican super-heavy crude oils with °API of 6.39 and 7.13, respectively, was studied.

The six ILs showed demulsifying effect, being family 1 the most efficient, achieving 95% of water removal from the heavy crude oils.

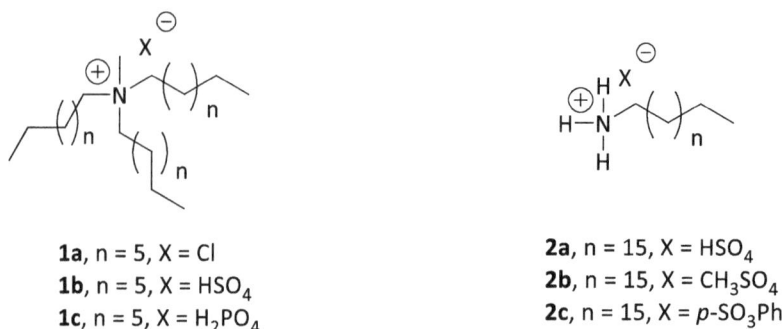

1a, n = 5, X = Cl
1b, n = 5, X = HSO₄
1c, n = 5, X = H₂PO₄

2a, n = 15, X = HSO₄
2b, n = 15, X = CH₃SO₄
2c, n = 15, X = *p*-SO₃Ph

Fig. (4.4). Structure of the ILs employed in the crude-oil-dehydration study by Flores *et al.*, 2014.

Through theoretical calculations, the quantum parameters were obtained at the semi-empirical levels RM1 and DFT as VM, MR, S and ω and allowed to establish a correlation with the experimental results and the data reported in the literature on ILs employed as demulsifying agents; from this study, the authors concluded that a good demulsifier has to cover some requisites such as high molecular volume, high polarization value for the cations and low molecular refraction values and molecular volume for the anions.

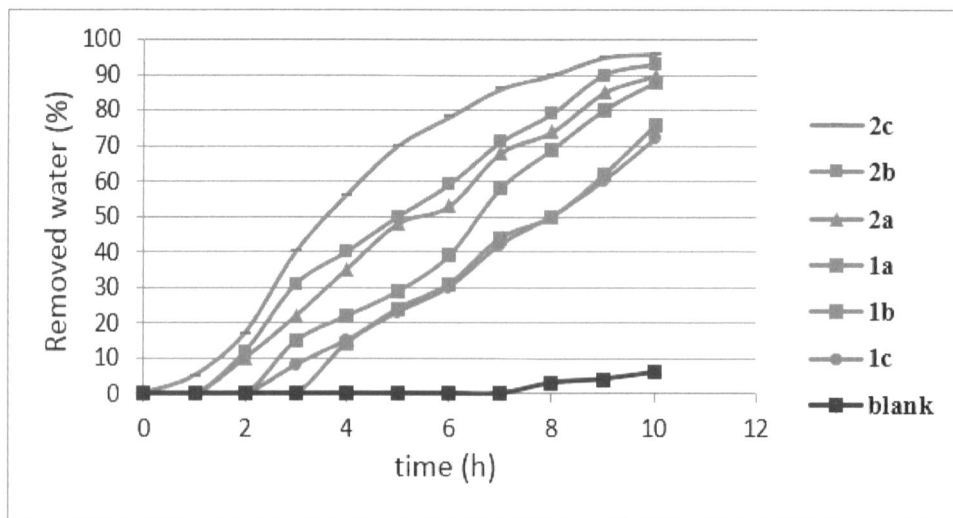

Fig. (4.5). Demulsification kinetics of the ILs synthesized in the work mentioned above, evaluated through the bottle test at 80°C and employing 500 ppm of demulsifiers.

In this sense, the article Demulsification of Water-in-Heavy Crude Oil Emulsion using Amphiphilic Ammonium Salts as Demulsifiers by the IMP researchers: Juan de la Cruz Clavel, Juan C. Navarro, Rafael Martínez-Palou. Tenside Surf.

Deterg. 2017, 54, 361-364 [42], describes the synthesis of six ILs with ammonium-type cation with different-length alkyl chains, which allowed to study the effect of the alkyl chain length on the efficiency of these compounds as demulsifying agents of water-in-oil (W/O) emulsions employing Mexican heavy crude oil. In this research work, a new methodology was used, which was faster and more accurate to evaluate the efficiency of the demulsifiers based on centrifugation-assisted demulsification. The proposed methodology was validated by means of the classical "bottle test". *N*-methyltrioctylammonium was the most efficient compound breaking the O/W emulsion as shown in Table **4.3** and confirmed through the bottle test, which showed the same trend, although the emulsion breaking times were higher (Fig. **4.5**).

Table 4.3. Results of the tests by centrifugation demulsification with ILs evaluated at different concentrations.

Demulsifier Chemical Structure	Demulsifier Concentration (ppm)	Centrifugation Time (min)					
		20	40	60	80	100	120
		Amount of Removed Water (%)					
Reference (without additive)	0	0	0	0	4	17	27
1a, n = 3	250	6	19	44	54	60	71
	500	10	20	48	57	64	76
1b, n = 9	250	0	13	24	40	50	65
	500	5	17	26	45	56	70
1c, n = 15	250	0	6	19	25	40	57
	500	0	10	22	36	50	63
2a, n = 1	250	7	19	43	57	78	71
	500	10	22	49	69	85	78

(Table 4.3) cont.....

Demulsifier Chemical Structure	Demulsifier Concentration (ppm)	Centrifugation Time (min)					
		20	40	60	80	100	120
		Amount of Removed Water (%)					
 2b, n = 3	250	8	37	51	61	84	90
	500	15	50	65	85	89	92
 2c, n = 6	250	12	40	55	68	90	93
	500	17	52	69	93	94	94

The alkylation reaction to obtain the compounds synthesized in this work was carried out by unconventional synthesis methods; in the case of family 1, the microreactor Q-tube was employed [43] and as for family 2, it was synthesized by the microwave device Discover de CEM [44].

The demulsification of W/O emulsions using ILs in combination with microwaves as Energy source was also demonstrated by a research group directed by Fortuny [45]. Some poly(ILs) have also shown to be efficient breaking water-in-oil emulsions [46, 47].

CONCLUDING REMARKS

As we have seen in the present chapters, emulsions cause operational problems in the oil industry, so the procedures for breaking them are of great importance in research in this area. ILs have demonstrated very good properties for breaking oil/water emulsions. ILs containing ammonium groups and long alkyl chain are very attractive ILs for demulsification by their amphiphilic properties, chemical stability, great commercial availability, low toxicity, and easy synthesis. Since the composition of crude oils can vary considerably from one region to another, the performance of ILs as demulsifiers can also be variable. The combined use of ILs with other physical methods such as microwaves and ultrasound allows faster and more effective demulsification in most cases.

REFERENCES

[1]   Abed, S.M.; Abdurahman, N.H.; Yunus, R.M.; Abdulbari, H.A.; Akbari, S. Oil emulsions and the different recent demulsification techniques in the petroleum industry - A review. *IOP Conf. Serie: Mater. Sci. Eng.*, **2019**.

[http://dx.doi.org/10.1088/1757-899X/702/1/012060]

[2]   Zolfaghari, R.; Fakhru'l-Razi, A.; Abdullah, L.C.; Elnashaie, S.S.E.H.; Pendashteh, A. Demulsification techniques of water-in-oil and oil-in-water emulsions in petroleum industry. *Separ. Purif. Tech.*, **2016**, *170*, 377-407.
[http://dx.doi.org/10.1016/j.seppur.2016.06.026]

[3]   Schramm, L.L. Petroleum Emulsion. In: *Emulsion Fundamentals and Applications in the Petroleum Industry*; American Chemical Society: Washington, DC, **1992**; pp. 1-45.

[4]   Martínez-Palou, R.; Aburto, J. Ionic liquids as Surfactants. Applications as demulsifiers of petroleum emulsions. In: *Ionic Liquids*; Current State of the Art. Intech, **2015**; pp. 305-326.
[http://dx.doi.org/10.5772/59094]

[5]   Salager, J.L. Microemulsions, en Handbook of Detergents – part A: Properties.Surfactant Science Series , **1999**; 82, pp. 253-302.

[6]   Alvarez, G.; Poteau, S.; Argillier, J.F.; Langevin, D.; Salager, J.L. Heavy Oil−Water Interfacial Properties and Emulsion Stability: Influence of Dilution. *Energy Fuels,* **2009**, *23*(1), 294-299.
[http://dx.doi.org/10.1021/ef800545k]

[7]   Kokai, S.L. *Crude Oil Emulsion. Petroleum Engineering Handbook*; SPE: Richardson, TX, **2005**.

[8]   Bera, A.; Mandal, A.; Guha, B.B. Synergistic effect of surfactant and salt mixture on interfacial tension reduction between crude oil and water in enhanced oil recovery. *J. Chem. Eng. Data,* **2014**, *59*(1), 89-96.
[http://dx.doi.org/10.1021/je400850c]

[9]   Abdurahman, N.H.; Rosli, Y.M.; Azhari, N.H.; Hayder, B.A. Pipeline transportation of viscous crudes as concentrated oil-in-water emulsions. *J. Petrol. Sci. Eng.,* **2012**, *90-91*, 139-144.
[http://dx.doi.org/10.1016/j.petrol.2012.04.025]

[10]   Chen, C.M.; Lu, C.H.; Chang, C-H.; Yang, Y.M.; Maa, J.R. Influence of pH on the stability of oil-i--water emulsions stabilized by a splittable surfactant. *Colloids Surf. A Physicochem. Eng. Asp.,* **2000**, *170*(2-3), 173-179.
[http://dx.doi.org/10.1016/S0927-7757(00)00480-5]

[11]   Pal, R. Effect of droplet size on the rheology of emulsions. *AIChE J.,* **1996**, *42*(11), 3181-3190.
[http://dx.doi.org/10.1002/aic.690421119]

[12]   Borges, B.; Rondón, M.; Sereno, O.; Asuaje, J. Breaking of water-in-crude-oil emulsions. 3. Influence of salinity and water-oil ratio on demulsifier action. *Energy Fuels,* **2009**, *23*(3), 1568-1574.
[http://dx.doi.org/10.1021/ef8008822]

[13]   Ahmed, N.S.; Nassar, A.M.; Zaki, N.N.; Gharieb, H.K. Stability and rheology of heavy crude oil-i--water emulsion stabilized by an anionic-nonionic surfactant mixture. *Petrol. Sci. Technol.,* **1999**, *17*(5-6), 553-576.
[http://dx.doi.org/10.1080/10916469908949734]

[14]   Spiecker, P.M.; Gawrys, K.L.; Trail, C.B.; Kilpatrick, P.K. Effects of petroleum resins on asphaltene aggregation and water-in-oil emulsion formation. *Colloids Surf. A Physicochem. Eng. Asp.,* **2003**, *220*(1-3), 9-27.
[http://dx.doi.org/10.1016/S0927-7757(03)00079-7]

[15]   Schorling, P.C.; Kessel, D.G.; Rahimian, I. Influence of the crude oil resin/asphaltene ratio on the stability of oil/water emulsions. *Colloids Surf. A Physicochem. Eng. Asp.,* **1999**, *152*(1-2), 95-102.
[http://dx.doi.org/10.1016/S0927-7757(98)00686-4]

[16]   Sullivan, A.P.; Kilpatrick, P.K. The effects of inorganic solid particles on water and crude oil emulsion stability. *Ind. Eng. Chem. Res.,* **2002**, *41*(14), 3389-3404.
[http://dx.doi.org/10.1021/ie010927n]

[17]   Maia Filho, D.C.; Ramalho, J.B.V.S.; Spinelli, L.S.; Lucas, E.F. Aging of water-in-crude oil

emulsions: Effect on water content, droplet size distribution, dynamic viscosity and stability. *Colloids Surf. A Physicochem. Eng. Asp.,* **2012**, *396*, 208-212.
[http://dx.doi.org/10.1016/j.colsurfa.2011.12.076]

[18]   Kokal, S. Society of Petroleum Enginering, Paper 77497. **2005**.

[19]   Salager, J-L.; Antón, R.E.; Forgiarini, A.; Márquez, L. Formulation of Microemulsions. In: *Microemulsions*; Stubenrauch, C., Ed.; WILEY-VCH Verlag GmbH & Co.: Berlin, Germany, **2009**; pp. 84-121.
[http://dx.doi.org/10.1002/9781444305524.ch3]

[20]   Ezzat, A.O.; Atta, A.M.; Al-Lohedan, H.A.; Aldalbahi, A. New amphiphilic pyridinium ionic liquids for demulsification of water Arabic heavy crude oil emulsions. *J. Mol. Liq.,* **2020**, *312*, 113407.
[http://dx.doi.org/10.1016/j.molliq.2020.113407]

[21]   Hazrati, N.; Miran Beigi, A.A.; Abdouss, M. Demulsification of water in crude oil emulsion using long chain imidazolium ionic liquids and optimization of parameters. *Fuel,* **2018**, *229*, 126-134.
[http://dx.doi.org/10.1016/j.fuel.2018.05.010]

[22]   Hezave, A.Z.; Dorostkar, S.; Ayatollahi, S.; Nabipour, M.; Hemmateenejad, B. Investigating the effect of ionic liquid (1-dodecyl-3-methylimidazolium chloride ([C12mim] [Cl])) on the water/oil interfacial tension as a novel surfactant. *Colloids Surf. A Physicochem. Eng. Asp.,* **2013**, *421*, 63-71.
[http://dx.doi.org/10.1016/j.colsurfa.2012.12.008]

[23]   Biniaz, P.; Farsi, M.; Rahimpour, M.R. Demulsification of water in oil emulsion using ionic liquids: Statistical modeling and optimization. *Fuel,* **2016**, *184*, 325-333.
[http://dx.doi.org/10.1016/j.fuel.2016.06.093]

[24]   Atta, A.M.; Al-Lohedan, H.A.; Abdullah, M.M.S.; ElSaeed, S.M. Application of new amphiphilic ionic liquid based on ethoxylated octadecylammonium tosylate as demulsifier and petroleum crude oil spill dispersant. *J. Ind. Eng. Chem.,* **2016**, *33*, 122-130.
[http://dx.doi.org/10.1016/j.jiec.2015.09.028]

[25]   Guzmán-Lucero, D.; Flores, P.; Rojo, T.; Martínez-Palou, R. Evaluation of ionic liquids as desemulsifier of water-in-crude oil emulsions. Study of microwave effect. *Energy Fuels,* **2010**, *24*(6), 3610-3615.
[http://dx.doi.org/10.1021/ef100232f]

[26]   Fang, C.S.; Lai, P.M.C. Microwave-heating and separation of water in-oil emulsions. *J. Microw. Power Electromagn. Energy,* **1995**, *30*(1), 46-57.
[http://dx.doi.org/10.1080/08327823.1995.11688257]

[27]   Nour, A.H.; Yunus, R.M. Stability and demulsification of water-in crude oil (w/o) emulsions *via* microwave heating. *J. Appl. Sci.,* **2006**, *6*, 1698-1702.
[http://dx.doi.org/10.3923/jas.2006.1698.1702]

[28]   Nour, A.H.; Yunus, R.M. A continuous microwave heating of water-in-oil emulsions: An experimental study. *J. Appl. Sci.,* **2006**, *6*, 1868-1872.
[http://dx.doi.org/10.3923/jas.2006.1698.1702]

[29]   Xia, L.X.; Lu, S.W.; Cao, G.Y. Salt-assisted microwave demulsification. *Chem. Eng. Commun.,* **2004**, *191*(8), 1053-1063.
[http://dx.doi.org/10.1080/00986440490276380]

[30]   Fortuny, M.; Oliveira, C.B.Z.; Melo, R.L.F.V.; Nele, M.; Coutinho, R.C.C.; Santos, A.F. Effect of salinity, temperature, water content, and pH on the microwave demulsification of crude oil emulsions. *Energy Fuels,* **2007**, *21*(3), 1358-1364.
[http://dx.doi.org/10.1021/ef0603885]

[31]   Holtze, C.; Sivaramakrishnan, R.; Antonietti, M.; Tsuwi, J.; Kremer, F.; Kramer, K.D. The microwave absorption of emulsions containing aqueous micro- and nanodroplets: A means to optimize microwave heating. *J. Colloid Interface Sci.,* **2006**, *302*(2), 651-657.

[http://dx.doi.org/10.1016/j.jcis.2006.07.020] [PMID: 16930614]

[32]   Mutyala, S.; Fairbridge, C.; Paré, J.R.J.; Bélanger, J.M.R.; Ng, S.; Hawkins, R. Microwave applications to oil sands and petroleum: A review. *Fuel Process. Technol.,* **2010,** *91*(2), 127-135. [http://dx.doi.org/10.1016/j.fuproc.2009.09.009]

[33]   Li, H.; Zhao, Z.; Xiouras, C.; Stefanidis, G.D.; Li, X.; Gao, X. Fundamentals and applications of microwave heating to chemicals separation processes. *Renew. Sustain. Energy Rev.,* **2019,** *114,* 109316. [http://dx.doi.org/10.1016/j.rser.2019.109316]

[34]   Santos, D.; da Rocha, E.C.L.; Santos, R.L.M.; Cancelas, A.J.; Franceschi, E.; Santos, A.F.; Fortuny, M.; Dariva, C. Demulsification of water-in-crude oil emulsions using single mode and multimode microwave irradiation. *Separ. Purif. Tech.,* **2017,** *189,* 347-356. [http://dx.doi.org/10.1016/j.seppur.2017.08.028]

[35]   da Silva, E.B.; Santos, D.; de Brito, M.P.; Guimarães, R.C.L.; Ferreira, B.M.S.; Freitas, L.S.; de Campos, M.C.V.; Franceschi, E.; Dariva, C.; Santos, A.F.; Fortuny, M. Microwave demulsification of heavy crude oil emulsions: Analysis of acid species recovered in the aqueous phase. *Fuel,* **2014,** *128,* 141-147. [http://dx.doi.org/10.1016/j.fuel.2014.02.076]

[36]   Abdurahman, N.H.; Yunus, R.M.; Azhari, N.H.; Said, N.; Hassan, Z. The Potential of Microwave Heating in Separating Water-in-Oil (w/o) Emulsions. *Energy Procedia,* **2017,** *138,* 1023-1028. [http://dx.doi.org/10.1016/j.egypro.2017.10.123]

[37]   Martínez-Palou, R. Applications of microwave for breaking petroleum emulsions. *Curr. Microw. Chem.,* **2017,** *4,* 276.

[38]   Dukhin, A.S.; Goetz, P.J. Acoustic and electroacoustic spectroscopy for characterizing concentrated dispersions and emulsions. *Adv. Colloid Interface Sci.,* **2001,** *92*(1-3), 73-132. [http://dx.doi.org/10.1016/S0001-8686(00)00035-X] [PMID: 11583299]

[39]   Binner, E.R.; Robinson, J.P.; Silvester, S.A.; Kingman, S.W.; Lester, E.H. Investigation into the mechanisms by which microwave heating enhances separation of water-in-oil emulsions. *Fuel,* **2014,** *116,* 516-521. [http://dx.doi.org/10.1016/j.fuel.2013.08.042]

[40]   Aburto, J.; Márquez, D.M.; Navarro, J.C.; Martínez-Palou, R. Amphiphilic Choline carboxilates Ionic Liquids as Demulsifiers of Water-in-Crude oil Emulsions. *Tenside Surf. Deterg,* **2014,** *51,* 314-317.

[41]   Flores, C.A.; Flores, E.A.; Hernández, E.; Castro, L.V.; García, A.; Alvarez, F.; Vázquez, F.S. Anion and cation effects of ionic liquids and ammonium salts evaluated as dehydrating agents for super-heavy crude oil: Experimental and theoretical points of view. *J. Mol. Liq.,* **2014,** *196,* 249-257. [http://dx.doi.org/10.1016/j.molliq.2014.03.044]

[42]   Clavel, J.C.; Navarro, J.C.; Martínez-Palou, R. Demulsification of Water-in-Heavy Crude Oil Emulsion using Amphiphilic Ammonium Salts as Demulsifiers. *Tenside Surfactants Deterg.,* **2017,** *54*(4), 361-364. [http://dx.doi.org/10.3139/113.110510]

[43]   http://www.qlabtech.com

[44]   http://www.cem.com

[45]   Lemos, R.C.B.; da Silva, E.B.; dos Santos, A.; Guimarães, R.C.L.; Ferreira, B.M.S.; Guarnieri, R.A.; Dariva, C.; Franceschi, E.; Santos, A.F.; Fortuny, M. Demulsification of water-in-crude oil emulsions using ionic liquids and microwave irradiation. *Energy Fuels,* **2010,** *24*(8), 4439-4444. [http://dx.doi.org/10.1021/ef100425v]

[46]   Atta, A.M.; Al-Lohedan, H.A.; Abdullah, M.M.S. Dipoles poly(ionic liquids) based on 2-acrylamid--2-methylpropane sulfonic acid-co-hydroxyethyl methacrylate for demulsification of crude oil water emulsions. *J. Mol. Liq.,* **2016,** *222,* 680-690.

[http://dx.doi.org/10.1016/j.molliq.2016.07.114]

[47]   Ezzat, A.O.; Atta, A.M.; Al-Lohedan, H.A.; Hashem, A.I. Synthesis and application of new surface active poly (ionic liquids) based on 1,3-dialkylimidazolium as demulsifiers for heavy petroleum crude oil emulsions. *J. Mol. Liq.,* **2018**, *251*, 201-211.
        [http://dx.doi.org/10.1016/j.molliq.2017.12.081]

CHAPTER 5

Application of ILs in the Transport of Heavy and Extra-heavy Crude Oils

Abstract: In the oil industry, in general, and in Mexico, in particular, the use of heavy crudes in the refining process has increased considerably. The processing of these heavy crudes implies an important technological challenge, and their transportation can sometimes be complicated. Among the alternatives for transporting this type of crude is the formation of invert emulsions in which ILs can play an important role, which is discussed in this chapter.

Keywords: Ionic liquids, Heavy-oil, Extra heavy-oil, Asphaltenes, Resins, API, Viscosity, Demulsifiers, Surfactants.

INTRODUCTION

Currently, most crude oil extracted from wells in Mexico, and in many other countries, is heavy oil (density equal to or below 20 ° API). The complex composition of these crude oils makes them difficult and expensive to be produced and transported through pipelines due to their low mobility and flow capacity, which are the result of high viscosity and specific gravity [1].

Heavy oil has been defined as oil with API gravity below 20°, which means that its relative density is above 0.933. It is any type of crude oil that does not flow easily. Density is usually defined in terms of API degrees (American Petroleum Institute) and is related to the specific gravity, the denser the oil, the lower the API density. The API densities of the liquid hydrocarbon range from 4° for bitumen rich in pitch up to 70° for condensates [2].

In general, heavy oil has attracted less attention as an energy resource due to the difficulties and costs associated with its production. It has been stated that there are more than 6 trillion barrels [1 trillion m^3] of oil in place (OIP) attributed to the heaviest hydrocarbons, equivalent to triple of the combined conventional oil and gas reserves in the world, thus deserving more careful attention [3].

Natural heavy crude oils display a wide density and viscosity spectrum. The viscosity at reservoir temperature is in general the most important measurement

Rafael Martínez Palou & Natalya V. Likhanova

for a hydrocarbon producer because it determines how easily oil will flow. Density is more important for oil refining, for it is a better indicator of the distillation derivatives. Unfortunately, there is not a clear correlation between both parameters. Intermediate or low-density crude oil with high paraffin content in a cold and shallow reservoir can have a higher viscosity than heavy crude oil, free of paraffin in a deep reservoir with high temperature. The production, transport and refining of heavy crude oil present special problems in comparison with light crude oil [4].

The physical properties that distinguish the heavy crude oils from the light ones include higher viscosity and density and the molecular weight comparison. Extra-heavy crude oil can even display viscosities above 10.000 centipoise (10 Pa·s) and 10° in the API index.

The oil density varies scarcely with temperature and for this reason, it has become the most used standard parameter in the oil industry to classify crude oils. Table **5.1** shows the classification of crude oils according to their densities.

Table 5.1. Classification of oils according to their API densities.

Crude Oil	Density (°API)
Extra-heavy	<10.0
Heavy	10.0 – 22.3
Intermediate	22.4 – 31.1
Light	31.2 – 39
Superlight	> 39.0

The reduction of the mobility rate is the main goal of the non-thermal recovery methods either by diminishing the oil viscosity or by increasing the viscosity of the displacing fluid.

The viscosity of fuels is correlated with the material average molecular weight, and it provokes frequently plugging problems both during oil extraction and transport. Examples of this are the Mexican crude oils "Maya" and "Ku-Maalob", which present compact asphaltenic aggregations whose values range in the interval from 6000 to 50,000 cp whereas the density is found between 22 and 8° API. An additional disadvantage is that the reservoir is far from the refinery and for this reason, a treatment enabling the crude flow through the country pipeline system is needed. The proposed solutions for transporting heavy and extra-heavy crude oil have been focused on the reduction of pressure drops in the transport systems originated by the increase in the fluid viscosity and friction generated between the fluid and pipe.

The high gravity of the crudes is generally due to the presence of bulky and complex molecules with high aromaticity such as asphaltenes and resins [5].

Asphaltenes are polyaromatic and polycyclic molecules that have heteroatoms and metals and represent the non-volatile-polar fraction of oil that is insoluble in *n*-alkanes. Asphaltenes appear surrounded and stabilized by resins and it is known that they have an electric charge and form aggregates by supramolecular attraction among them and tend to precipitate due to pressure, temperature and oil composition changes, which generate the plugging of pipelines and evident increase in the oil's viscosity [6 - 9].

When isolated, asphaltenes are dark brown to black molecules; they form amorphous powder with specific gravity above the unit; their average molecular weight oscillates between 750 and 10000 g/mol and do not have a defined melting point, but decompose at temperatures between 300 and 400°C or even higher [10]. The IMP has made every effort in the research of the asphaltene structures [11], which has helped importantly to elucidate them and understand the electron distribution and aromaticity, which have been explained through the rule developed by Dr. Ruiz-Morales [12 - 16].

Chemically, the asphaltene structures consist of aromatic and aliphatic polycyclic systems with the presence of naphthenic acids, heterocycles, substituted phenols and carboxylic acids and contain metals such as nickel, vanadium and iron. The asphaltene structures are not well defined, but several possible structures have been proposed to explain the composition and properties of the asphaltene fraction. A hypothetical structure of the asphaltenes is shown in Fig. (**5.1**).

Resins are defined as the non-volatile-polar fraction of crude oil that is soluble in *n*-pentane, *n*-heptane, and aromatic solvents such as toluene, but insoluble in methanol and propanol. Resins are also dark brown to black molecules and are semisolid compounds with specific gravity close to unity and their molecular weight ranges from 500 to 2000 g/mol, being highly adhesive materials. The resin content in crude oils is between 2 and 40 wt.%. The molecular species are the same as those in aromatic compounds, but in resins, they have both higher molecular weight and polarity, presence of heteroatoms and lower hydrogen-carbon rate than that of aromatic compounds. The resin fraction consists of carbon, hydrogen, oxygen, nitrogen, and naphthenic acids.

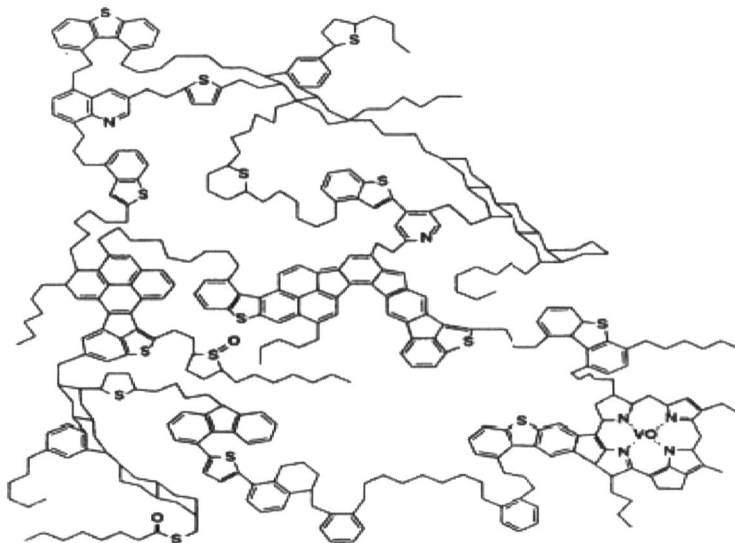

Fig. (5.1). Schematic representation of an asphaltene molecule.

Chemically, oil resins present polyaromatic and polycyclic systems with highly polar groups at the far ends with heteroatoms such as sulfur, oxygen or nitrogen including pyrrole and indole groups; also, non-polar-long-chain groups are to be found. Fig. (**5.2**) shows a hypothetical structure of a resin molecule.

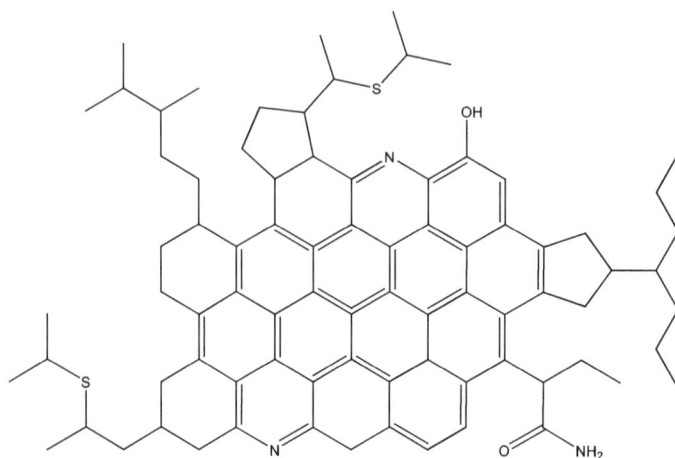

Fig. (5.2). Schematic representation of a resin molecule.

Different technological alternatives have been studied to improve the transport properties of heavy crudes; on some occasions, it is necessary to choose more than one option to enhance the transport of heavy crude oils, for any technology by itself is enough to obtain the desired results, because the crude physical characteristics and available transport systems determine the selection and combination of the alternatives to be employed.

In the review Transportation of Heavy and Extra-heavy Crude Oil by Pipelines: A review by Martínez-Palou, R., Mosqueira, M. L., Zapata-Rendón, B., Mar-Juárez, E., Bernal-Huicochea, C., Cruz Clavel, J. and Aburto, J. J. Proc. Sci. Eng. 2011, 75, 274-282 [17], different strategies employed to solve the transport of heavy and extra-heavy crude oils, which are mainly focused in the reduction of viscosity and friction are discussed as shown in Fig. (**5.3**).

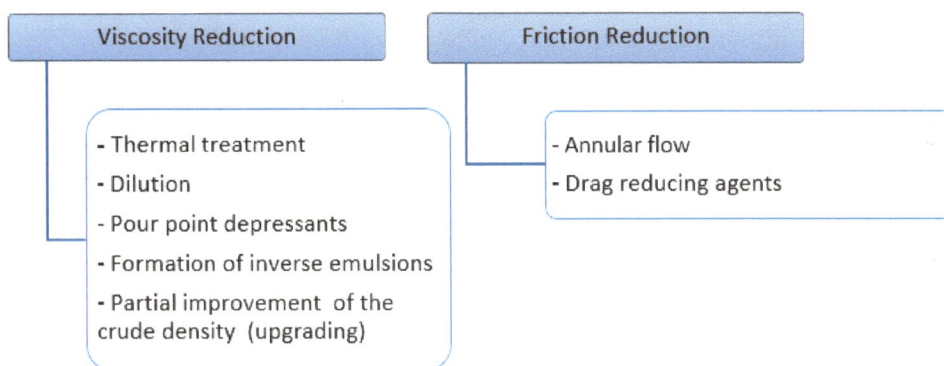

Fig. (5.3). Some strategies aimed at improving the transport of heavy crude oils.

These strategies are explained briefly as follows:

Thermal treatment (Heating). The crude density is reduced proportionally to the temperature increase; for this reason, crude heating is one of the most popular alternatives for transport, however, it implies high energy consumption and the crude production costs also become higher.

Dilution. It is a technology that involves the dilution of crude oil with solvents such as hexane obtained from its very refining and through the addition of condensates from the production of natural gas and in some cases of naphtha, kerosene or light oil at 20-30% of crude oil with which the crude viscosity is improved significantly and facilitates further processes such as desalting and dehydration. One of the drawbacks of this technology is the incorporation of previously processed products; on the other hand, this procedure affects the equilibrium of asphaltenes in such a way that precipitation or separation of the solid phase can occur inside the transmission pipelines and storing tanks, provoking obstructions that make difficult the transport, and cause overpressure [18, 19].

Partial improvement of crude oil (Upgrading). *Upgrading* is a technology that is very attractive, for it facilitates the transport by improving the flow properties of crude oil in the production wells by physicochemical methods that in general combine the use of hydrogen and catalysts, heating and high pressures. This process, known as hydroprocessing, can include processes such as hydrocracking, hydrotreatment and hydrorefining, which allow to modify remarkably the crude viscosity, properties and chemical composition. The IMP has devoted every effort to do research to develop the technology developed by Dr. Ancheyta *et al* [20 - 33].

Annular flow. The annular flow regime can be achieved by mixing two immiscible liquids such as water and crude oil with very different viscosities in a horizontal pipe. This strategy is intended to improve the flow and pumping stability through the formation of either a water or solvent film around the pipe walls to work as a lubricant, for the heavy crude oil is concentrated in the pipe center; such action reduces the pressure and induces the crude oil to flow (Fig. 5.4).

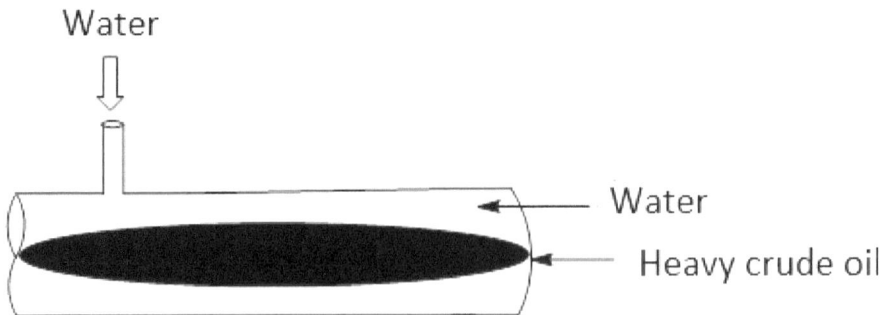

Fig. (5.4). Transport of heavy crude oil by means of annular flow.

This procedure allows to reach suction pressures that are very similar to those of water; the required solvent volume oscillates between 10 and 30%, depending on the characteristics of the crude oil to be transported. The main drawback of this option is the difficulty to keep the right annular flow and crude oil in the pipe center and water around the walls, in addition to the important amount of water that is needed, which has to be afterward removed from crude oil and that on some occasions can form an emulsion with it [34].

Drag reducing agents (DRAs). The role played by these additives is to suppress the growth of turbulent whirlpools that occur commonly during the transport of crude oils, which results in higher flow with pumping at constant pressure. Through the addition of chemical compounds such as certain surfactants or polymers like low-density polyethylene, criss-cross-linear-alpha-olefin copolymers, polyacrylamides and polymethacrylates, it is possible to reduce the friction close to the pipe walls and inside the core of the turbulent flow of the moving fluid [35].

Pour point depressants. The lowest temperature at which oil can be poured is known as the pour point. Pour point depressants are chemical products that are alternatively known as wax crystal modifiers, which are capable of modifying nucleation, adsorption or wax solubility. Wax crystal modifiers are, in general, polymers (branched-poly-α-olefins, fatty-acid amides, poly-*n*-alkyl acrylates and methacrylate copolymers, among others) that help depress the crude oil pour point, viscosity and yield stress remarkably, which facilitates the transport of crude oils with high wax contents [36 - 42]. In this context, the IMP has evaluated some copolymers and terpolymers based on combinations of vinyl, styrene and *n*-burylacrylate acetates, which work as viscosity reducers and modify the rheological properties of the crude oil, improving the flow of Mexican heavy crude oils [43, 44].

Formation of inverse emulsions. This strategy consists of obtaining an inverse emulsion, *i.e.* one where water is the continuous phase and crude oil is the dispersed phase by means of an emulsifier agent, which is represented as O/W. The emulsion helps diminish the crude density and eases its fluidity and transport. Once crude oil is transported, it is necessary to break the emulsion and separate water before it is sent to the refining process.

An emulsion is generally defined as a system where a liquid is distributed or relatively dispersed in the form of droplets within another substantially immiscible liquid. Emulsions are of practical interest due to their presence in everyday life. They can be found in important areas such as food, cosmetics, paste and paper, pharmaceutical and agricultural products. Oil emulsions consist of water emulsified within the crude oil and are known as direct emulsions, which are the most commonly found in crude oil and less common are the inverse emulsions (oil in water), but also complex or mixed emulsions can occur. Crude emulsions are undesirable, for they produce high pumping costs, pipe corrosion and require the use of special equipment. Emulsions can be found throughout all the oil recovery stages and in the transformation industry (drilling fluid, production, processing plant and transport of emulsions) [45, 46].

The formation of O/W emulsions by means of the industrial process known as Orimulsion, which was employed in Venezuela by the end of the past century, developing a fuel emulsion by dispersing in water bitumen from the Orinoco Oil Belt Orinoco [47 - 50], has shown to be a very effective method because it is relatively cheap and easy to be applied industrially, however, this method has the disadvantage that after the transport of crude oil, it remains emulsified and for this reason, a very efficient demulsifying and dehydrating method is required before proceeding to processing and refining.

In this context, the IMP has made every effort to do research aimed at developing a transport technology based on the formation of inverse emulsions.

In the work, Study on the formation and breaking of extra heavy crude oil-i--water emulsions: A proposal strategy for transportation of extra heavy crude oils by Martínez-Palou, R., Reyes, J., Cerón, R., Villanueva, D., Vallejo, A. A., and Aburto, J. Chem. Eng. Proc. Proc. Intensif. 2015, 98, 112-122 [51], an integral strategy for the transport of heavy crude oils through the selection of the most suitable emulsifier from several commercial or synthetic products is discussed, finding that a mixture of the commercial product of the nonylphenol-ethoxylate (TQA) type and a polyglucoside with 12-carbon alkyl chain (H4C12) was the best emulsifier to form a stable inverse emulsion from a Mexican extra-heavy crude oil. On the other hand, the emulsion was characterized by calorimetry, optical microscopy, particle size (Fig. **5.5**) and dynamic viscosity and the stability of the emulsions was evaluated by the separation of water in the presence of surfactants and heating. Also, the evaluation of different surfactants synthesized in the IMP laboratories was carried out, finding that an ionic biosurfactant derived from glycine forming a 14-carbon ester was an environmentally friendly demulsifier and efficient, separating 89.5% of water from the emulsion. It is worth mentioning that at the beginning of the refining process, crude oil is submitted to a dehydrating and desalting process through which water can be more deeply removed from crude oil.

The surfactants employed in this integral procedure, which was previously commented, were synthesized in the IMP laboratories; in the case of the surfactant H4C12, it was produced by means of conventional and microwave methods, which were described in the research work Microwave-assisted Organic Synthesis *vs* Conventional Heating: A comparative study for Fisher Glycosidation of monosaccharides by R. Cerón-Camacho, J. Aburto, Luisa E. Montiel and Rafael Martínez-Palou. Comptes Rendus Chimie 2013, 16, 427-432 [52].

Likewise, ionic demulsifiers were synthesized according to the procedure described in the article Efficient Microwave-Assisted Synthesis of Ionic Esterified

Amino Acids by Cerón-Camacho, R., Aburto, J., Montiel, L. E.; Flores, Cuellar F., and Martínez-Palou, R. Molecules 2011, 16, 8733-8744 [53].

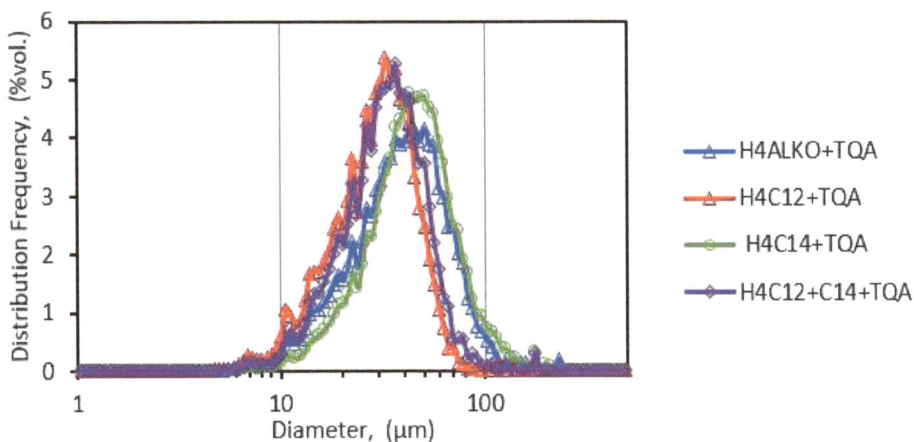

Fig. (5.5). Distribution of the particle size of some emulsions that were prepared with a mixture of commercial surfactant (TQA).

Looking for the best emulsifier prototypes, the products developed at the IMP were evaluated through different studies, *e.g.* in the case of emulsifiers, a synergistic effect was identified among surfactants of the types alkyl polyglucoside and cellobioside for the formation of emulsions, which was described in the article by Cerón-Camacho R, Martínez-Palou R, Chávez-Gómez B, Cuéllar F, Bernal-Huicochea C, de la Cruz Clavel J, and Aburto, J., titled Synergistic effect of alkyl-O-glucoside and cellobioside biosurfactants as effective emulsifiers of crude oil in water: A proposal for the transport of heavy crude oil by pipeline. Fuel 2013, 110, 310-317 [54]. Fig. (**5.6**) shows an image of the obtained inverse emulsion, where oil drops dispersed in water as a continuous medium (blue color) can be observed.

Other studies focused on the action way of demulsifiers and the optimization of the conditions to enhance their performance produced articles like the one titled Desemulsification of heavy crude oil in water emulsion: A comparative study between Microwave and Conventional Heating by Martínez-Palou, R., Cerón-Camacho, R.; Chávez, B.; Vallejo, A. A.; Aburto, J.; Villanueva, D.; and Reyes J. Fuel 2013, 113, 407-414 [55], where it was observed that the use of microwaves can be an alternative heating source that makes the emulsion breaking process more efficient with respect to time and amount of separated water, as it can be seen in Fig. (**5.7**).

Fig. (5.6). Optimal micrograph of the inverse emulsion obtained from heavy crude oil using emulsifiers developed at the IMP.

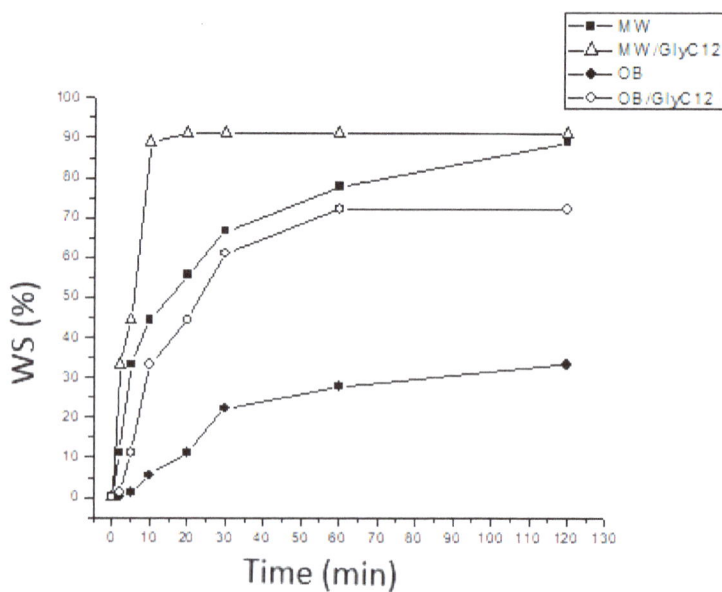

Fig. (5.7). Percentage of water separation (WS) in the emulsion containing the demulsifier GlyC12 (1000 ppm) after different heating times in oil bath (OB, ○) and microwaves (MW, Δ) and OB without demulsifier (●) and MW without demulsifier (■).

The IMP has also studied other alternatives different from ionic liquids (ILs) and chemical products for the breaking of inverse emulsions, which is the case of fungal spores with very good results; this topic was dealt with in the article Demulsification of crude oil-in-water emulsions by means of fungal spores by Vallejo, A., Martínez-Palou, R., Cerón, R., Chávez, B., Reyes, J. and Aburto, J. PLos One 2017, 1-17 [56].

Studies on the transport of heavy crude oils by forming emulsions through technology developed at the IMP resulted in inventions that were protected in the following patents:

ILs as viscosity reducers of heavy crude oils. Eugenio A. Flores, Rafael Martínez-Palou, Diego Guzmán, Natalya V. Likhanova. Canadian Pat. Appl. No. 2,708,416 (2010) [57].

Líquidos iónicos como reductores de viscosidad en crudos pesados (Ionic liquids as viscosity reducers of heavy crude oils). Rafael Martínez Palou, Eugenio Alejandro Flores Oropeza, Diego Javier Guzmán Lucero, Natalya V. Lykhanova (2009). Issue number: MX/A/2009/007078 [58].

Process for demulsification of crude oil in water emulsions by means of natural or synthetic amino acid-based demulsifiers. Jesús R. Reyes, R. Martínez-Palou, Ricardo Cerón, Alba A. Vallejo, R. Rodríguez, B. Chávez, Jorge Aburto. Canadian Pat. 2,852,865 (2016) [59].

Improved process for obtaining ionic amino acid esters. Rafael Martínez-Palou, Ricardo Cerón, Alba A. Vallejo, R. Jesús Reyes, C. Bernal, M. Ramirez de Santiago, Jorge Aburto. Canadian Pat. 2,903,564 (2018).

Process for Demulsification of Crude Oil in Water Emulsions by means natural or synthetic aminoacid-based demulsifiers. J. Reyes, R. Martínez-Palou, R. Cerón, Alba A. Vallejo, R. Rodriguez, B. Chavez, J. Aburto. US Pat. 9,677,009 B2 (2017).

Proceso mejorado para la obtención de éteres de aminoácidos (Improved process for the synthesis of amino acid esters). Rafael Martínez Palou, R. Ceron, A. A. Vallejo Cardona, J. Reyes, J. Clavel, M. Ramirez de Santiago, J. Aburto. Pat. Colombiana. Exp. CO 15-211.717. Application date: 08/09/2015. C7C 227/18 (2018) [60].

Process for Obtaining Amino Acid Esters. Rafael Martínez-Palou, R. Cerón, A. A. Vallejo, R. Jesús Reyes, C. Bernal, M. Ramirez de Santiago, J. Aburto. US Pat. 10,384,247 B2 (2019) [61].

The interaction of the different components of crude oils, mainly of heavy crude oils, with ILs is well known and has been discussed by different authors as part of transport alternatives for heavy oils, which is the case of the studies related to the inhibition of asphaltenes precipitation [62 - 66] conducted by the IMP researchers Murillo-Hernández, J. A.; and Aburto, J. in the work entitled Current Knowledge and Potential Applications of Ionic Liquids in the Petroleum Industry. Ionic Liquids: Applications and Perspectives, InTech, 2011. DOI: 10.5772/ 13974 [67]. In such chapter, a critical description of the state of the art of some applications of ILs in the oil industry up to 2009, year in which this book was published, was done.

The interactions between ILs and asphaltenes were studied experimentally through fluorescence spectroscopy by Murillo-Hernández, J.; García-Cruz, I.; López-Ramírez, S.; Durán-Valencia, C.; Domínguez, J.M.; and Aburto, J. in the work entitled Aggregation Behavior of Heavy Crude Oil-Ionic Liquid Solutions by Fluorescence Spectroscopy. Energy. Fuel 2009, 23, 4584-4592 [68]. In this work, the effect of ILs as inhibitors or aggregation stabilizers of asphaltenes was studied by quantum spectroscopy and spectrofluorometry.

The precipitation of asphaltenes is one of the factors that make difficult the transport of crude oil, produces pipe plugging and stops processes where the formation of asphaltenic aggregates is involved. Some electronic properties and critical aggregation concentration (CAC) have been characterized, finding that the presence of polyatomic anions such as $[AlCl_4]^-$, $[BF_4]^-$ and $[PF_6]^-$ increases the HOMO-LUMO (ΔH-L) separation, electronegativity (χ), absolute hardness (η) and dipolar moment (μ) in comparison with monatomic ILs like those containing $[Br]^-$. In addition, the CAC values of the ILs displayed a linear correlation with the dipolar moment. The CAC value was below 50 ppm, which according to the authors, indicates strong interactions between the IL and asphaltene; some ILs seem to form a protective layer around the asphaltene molecules, which prevents their agglomeration and precipitation. These effects were evidenced by the displacement observed in the mass spectral center in the fluorescence emission spectra of heavy crude oil with and without ILs.

The presence of ILs in a heptane and toluene solution of a heavy crude oil displaced the aggregation point value to a higher HCO concentration or *n*-heptane volume, probably because the ILs modified the relative polarity of the microenvironment around the HCO molecules and changed the HCO aggregation to higher values, specially by employing an IL with pyridinium as cation and PF_6^- as anion.

The interactions between the ILs and crude oil components have also been studied by means of theoretical computations by other IMP research groups, like that integrated by Hernandez-Bravo, R.; Miranda, A. D.; Martínez-Mora, O.; Domínguez, Z.; Martínez-Magadan, J. M.; García-Chávez, R.; and Domínguez-Esquivel, J. M. who in the work entitled Calculation of the Solubility Parameters by COSMO-RS Methods and its Influences on Asphaltenes-Ionic Liquids Interactions. Ind. Eng. Chem. Res. 2017, 56, 5107−5115 [69], presented a study to describe the solubility behavior of seventeen ILs with asphaltenes using the DFT approximation with the base def-TVZP and the density functional Perdew-Burk--Ernerhof (PBE).

Using the conductor COSMO-RS with the real solvent approximation, the density, molar volume, viscosity and caloric capacity of these compounds were established. The solubility parameters of the ILs and asphaltenes were determined through a relationship between solubility and density. The relative solubility of asphaltenes in the ILs was found using the parameter δ and the Hansen sphere method, giving very similar results to those obtained experimentally. A solubility order was established, where [EMIM]Cl was the most soluble and [BMIM]Cl was the least soluble. From this study, it was stated that the cation structure exerts a great effect on the solubility of ILs, which includes van der Waals, π-π, cation-π, and hydrophobic interactions. Some solubility values established by these authors are shown in Table **5.2**.

Table 5.2. Relative solubility of model asphaltenes in some ILs.

Models	RED
[EMIM]Cl	0.385
[EMIM]Tf$_2$N	1.504
[BMIM]Cl	1.079
[EMIM]Tf$_2$N	1.521
[C6MIM]Cl	0.399
[C6MIM]Tf$_2$N	1.247

In the same sense, in another work by Hernández-Bravo, R., Miranda, A. D., Martínez-Magadán, J. M., and Domínguez, J. M. titled Experimental and Theoretical Study on Supramolecular Ionic Liquid (IL)–Asphaltene Complex Interactions and Their Effects on the Flow Properties of Heavy Crude Oils. J. Phys. Chem. B 2018, 122, 4325–4335 [70], a theoretical-experimental study was developed in order to understand the molecular interactions between asphaltenes and ILs. The computational numerical simulation calculations using the density-functional theory (DFT) with dispersion corrections and molecular dynamics

(MD) confirmed the formation of supramolecular IL-asphaltene complexes, which modified the properties of crude oils such as viscosity and interfacial tension. The IL cation and π interactions with asphaltene seem to prevail in the IL-asphaltene interaction mechanism, which allowed to predict the experimental results observed through rheological studies of crude oil before and after being doped with different ILs. According to these results, the IL 2-tetradecylisoquinolinium bromide displayed the best effect in reducing the viscosity of heavy crude oils based on the calculated interaction energies.

The application of ionic liquids for breaking emulsions, with [71] and without [72], the aid of microwaves has recently been reviewed.

CONCLUDING REMARKS

As we have seen in the present chapters, with careful selection of the chemical structure, some ILs can be actual invert emulsion breakers used as a strategy for transporting heavy crude oil. Many ILs have shown that they can be used as emulsifying or demulsifying agents for oil emulsions, especially those that contain long alkyl chains that give them an amphiphilic character (polar zone in the ionic head and apolar zone in the hydrocarbon chain). Heavy oil upgrading technologies and the formation of inverse emulsions are two very convenient alternatives for transporting heavy crude oil.

REFERENCES

[1]   Speight, J.G. *Chemical and Technology of Petroleum*; Marcel Dekker: New York, **1991**.

[2]   Sjoblom, J.; Johnsen, E.E.; Westvik, A.; Ese, M.H.; Djuve, J.; Auflem, I.H.; Kallevik, H. Demulsifiers in the oil industry. In: *Encyclopedic Handbook of Emulsion Technology*; Sjoblom, J., Ed.; Marcel Dekker: New York, **2001**; p. 595.

[3]   Schramm, L.L. Petroleum Emulsion. In: *Emulsion Fundamentals and Applications in the Petroleum Industry*; American Chemical Society: Washington, DC, **1992**; pp. 1-45.

[4]   Gafonova, O.V. *Tesis de Maestría: Role of Asphaltenes and Resines in the Stabilization of Water-i--Hydrocarbon Emulsion*; The University of Calgary, **2000**.

[5]   Duran, J.A.; Casas, Y.A.; Xiang, L.; Zhang, L.; Zeng, H.; Yarranton, H.W. Nature of Asphaltene Aggregates. *Energy Fuels,* **2019**, *33*(5), 3694-3710.
[http://dx.doi.org/10.1021/acs.energyfuels.8b03057]

[6]   Bian, H.; Kan, A.; Yao, Z.; Duan, Z.; Zhang, H.; Zhang, S.; Zhu, L.; Xia, D. Impact of Functional Group Methylation on the Disaggregation Trend of Asphaltene: A Combined Experimental and Theoretical Study. *J. Phys. Chem. C,* **2019**, *123*(49), 29543-29555.
[http://dx.doi.org/10.1021/acs.jpcc.9b07695]

[7]   Glova, A.D.; Larin, S.V.; Nazarychev, V.M.; Kenny, J.M.; Lyulin, A.V.; Lyulin, S.V. Toward Predictive Molecular Dynamics Simulations of Asphaltenes in Toluene and Heptane. *ACS Omega,* **2019**, *4*(22), 20005-20014.
[http://dx.doi.org/10.1021/acsomega.9b02992] [PMID: 31788635]

[8]   Xue, H.; Wang, C.; Jiang, L.; Wang, H.; Lv, Z.; Huang, J.; Xiao, W. Asphaltene precipitation trend and controlling its deposition mechanism. *Nat. Gas Ind. B,* **2022**, *9*(1), 84-95.

[http://dx.doi.org/10.1016/j.ngib.2021.12.001]

[9]   Ok, S.; Mal, T.K. NMR Spectroscopy Analysis of Asphaltenes. *Energy Fuels,* **2019**, *33*(11), 10391-10414.
[http://dx.doi.org/10.1021/acs.energyfuels.9b02240]

[10]   Altgelt, K.; Boduszynski, M. *Composition and analysis of heavy petroleum fractions*; Marcel Dekker Inc.: New York, **1994**.

[11]   Buch, L.; Groenzin, H.; Buenrostro-Gonzalez, E.; Andersen, S.I.; Lira-Galeana, C.; Mullins, O.C. Effect of hydrotreatment on asphaltene fractions. *Fuel,* **2003**, *82*, 1075-1084.
[http://dx.doi.org/10.1016/S0016-2361(03)00006-1]

[12]   Ruiz-Morales, Y.; Mullins, O.C. Polycyclic Aromatic Hydrocarbons of Asphaltenes Analyzed by Molecular Orbital Calculations with Optical Spectroscopy. *Energy Fuels,* **2007**, *21*(1), 256-265.
[http://dx.doi.org/10.1021/ef060250m]

[13]   Ruiz-Morales, Y.; Wu, X.; Mullins, O.C. Electronic Absorption Edge of Crude Oils and Asphaltenes Analyzed by Molecular Orbital Calculations with Optical Spectroscopy. *Energy Fuels,* **2007**, *21*(2), 944-952.
[http://dx.doi.org/10.1021/ef0605605]

[14]   Ruiz-Morales, Y.; Mullins, O.C. Simulated and measured optical absorption spectra of asphaltenes. *Energy Fuels,* **2009**, *23*, 1169-1177.
[http://dx.doi.org/10.1021/ef800663w]

[15]   Ruiz-Morales, Y. Aromaticity in pericondensed cyclopenta-fused polycyclic aromatic hydrocarbons determined by density functional theory nucleus-independent chemical shifts and the Y-rule — Implications in oil asphaltene stability. *Can. J. Chem.,* **2009**, *87*(10), 1280-1295.
[http://dx.doi.org/10.1139/V09-052]

[16]   Ruiz-Morales, Y.; Miranda-Olvera, A.D.; Portales-Martínez, B.; Domínguez, J.M. Determination of [13] C NMR Chemical Shift Structural Ranges for Polycyclic Aromatic Hydrocarbons (PAHs) and PAHs in Asphaltenes: An Experimental and Theoretical Density Functional Theory Study. *Energy Fuels,* **2019**, *33*(9), 7950-7970.
[http://dx.doi.org/10.1021/acs.energyfuels.9b00182]

[17]   Martínez-Palou, R.; Mosqueira, M.L.; Zapata-Rendón, B.; Mar-Juárez, E.; Bernal-Huicochea, C.; de la Cruz Clavel-López, J.; Aburto, J. Transportation of heavy and extra-heavy crude oil by pipeline: A review. *J. Petrol. Sci. Eng.,* **2011**, *75*(3-4), 274-282.
[http://dx.doi.org/10.1016/j.petrol.2010.11.020]

[18]   Hart, A. A review of technologies for transporting heavy crude oil and bitumen *via* pipelines. *J. Pet. Explor. Prod. Technol.,* **2014**, *4*(3), 327-336.
[http://dx.doi.org/10.1007/s13202-013-0086-6]

[19]   Gateau, P.; Hénaut, I.; Barré, L.; Argillier, J.F. Heavy Oil Dilution. *Oil Gas Sci. Technol.,* **2004**, *59*(5), 503-509.
[http://dx.doi.org/10.2516/ogst:2004035]

[20]   Fukuyama, H.; Terai, S.; Uchida, M.; Cano, J.L.; Ancheyta, J. Active carbon catalyst for heavy oil upgrading. *Catal. Today,* **2004**, *98*(1-2), 207-215.
[http://dx.doi.org/10.1016/j.cattod.2004.07.054]

[21]   Sánchez, S.; Rodríguez, M.A.; Ancheyta, J. Kinetic model for moderate hydrocracking of heavy oils. *Ind. Eng. Chem. Res.,* **2005**, *44*(25), 9409-9413.
[http://dx.doi.org/10.1021/ie050202+]

[22]   Sámano, V.; Guerrero, F.; Ancheyta, J.; Trejo, F.; Díaz, J.A.I. A batch reactor study of the effect of deasphalting on hydrotreating of heavy oil. *Catal. Today,* **2010**, *150*(3-4), 264-271.
[http://dx.doi.org/10.1016/j.cattod.2009.09.004]

[23]   Martínez, J.; Ancheyta, J. Kinetic model for hydrocracking of heavy oil in a CSTR involving short

term catalyst deactivation. *Fuel,* **2012**, *100*, 193-199.
[http://dx.doi.org/10.1016/j.fuel.2012.05.032]

[24]   Ortiz-Moreno, H.; Ramírez, J.; Cuevas, R.; Marroquín, G.; Ancheyta, J. Heavy oil upgrading at moderate pressure using dispersed catalysts: Effects of temperature, pressure and catalytic precursor. *Fuel,* **2012**, *100*, 186-192.
[http://dx.doi.org/10.1016/j.fuel.2012.05.031]

[25]   Castañeda, L.C.; Muñoz, J.A.D.; Ancheyta, J. Combined process schemes for upgrading of heavy petroleum. *Fuel,* **2012**, *100*, 110-127.
[http://dx.doi.org/10.1016/j.fuel.2012.02.022]

[26]   Nguyen, T.S.; Tayakout-Fayolle, M.; Ropars, M.; Geantet, C. Hydroconversion of an atmospheric residue with a dispersed catalyst in a batch reactor: Kinetic modeling including vapor–liquid equilibrium. *Chem. Eng. Sci.,* **2013**, *94*, 214-223.
[http://dx.doi.org/10.1016/j.ces.2013.02.036]

[27]   Angeles, M.J.; Leyva, C.; Ancheyta, J.; Ramírez, S. A review of experimental procedures for heavy oil hydrocracking with dispersed catalyst. *Catal. Today,* **2014**, *220-222*, 274-294.
[http://dx.doi.org/10.1016/j.cattod.2013.08.016]

[28]   Calderón, C.J.; Ancheyta, J. Modeling of slurry-phase reactors for hydrocracking of heavy oils. *Energy Fuels,* **2016**, *30*(4), 2525-2543.
[http://dx.doi.org/10.1021/acs.energyfuels.5b02807]

[29]   Félix, G.; Quitian, A.; Rodríguez, E.; Ancheyta, J.; Trejo, F. Methods to calculate hydrogen consumption during hydrocracking experiments in batch reactors. *Energy Fuels,* **2017**, *31*(11), 11690-11697.
[http://dx.doi.org/10.1021/acs.energyfuels.7b01878]

[30]   Félix, G.; Ancheyta, J.; Trejo, F. Sensitivity analysis of kinetic parameters for heavy oil hydrocracking. *Fuel,* **2019**, *241*, 836-844.
[http://dx.doi.org/10.1016/j.fuel.2018.12.058]

[31]   Félix, G.; Ancheyta, J. Comparison of hydrocracking kinetic models based on SARA fractions obtained in slurry-phase reactor. *Fuel,* **2019**, *241*, 495-505.
[http://dx.doi.org/10.1016/j.fuel.2018.11.153]

[32]   *Hydroprocessing of heavy oils and residua,* 1st Ed.; CRC Press: Boca Raton, Florida, **2007**.

[33]   Ancheyta, J.; Sánchez, S.; Rodríguez, M.A. Kinetic modeling of hydrocracking of heavy oil fractions: A review. *Catal. Today,* **2005**, *109*(1-4), 76-92.
[http://dx.doi.org/10.1016/j.cattod.2005.08.015]

[34]   Ghosh, S.; Mandal, T.K.; Das, G.; Das, P.K. Review of oil water core annular flow. *Renew. Sustain. Energy Rev.,* **2009**, *13*(8), 1957-1965.
[http://dx.doi.org/10.1016/j.rser.2008.09.034]

[35]   Fink, J. *Drag Reduction and Flow Improvement in Guide to the Practical Use of Chemicals in Refineries and Pipelines*; Elsevier Inc, **2016**.

[36]   Langevin, D.; Poteau, S.; Hénaut, I.; Argillier, J.F. Crude oil emulsion properties and their application to heavy oil transportation. *Oil Gas Sci. Technol.,* **2004**, *59*(5), 511-521.
[http://dx.doi.org/10.2516/ogst:2004036]

[37]   Pedersen, K.S.; Rønningsen, H.P. Influence of wax inhibitors on wax appearance temperature, pour point, and viscosity of waxy crude oils. *Energy Fuels,* **2003**, *17*(2), 321-328.
[http://dx.doi.org/10.1021/ef020142+]

[38]   da Silva, C.X.; Álvares, D.R.S.; Lucas, E.F. New additives for the pour point reduction of petroleum middle distillates. *Energy Fuels,* **2004**, *18*(3), 599-604.
[http://dx.doi.org/10.1021/ef030132o]

[39]   Song, Y.; Ren, T.; Fu, X.; Xu, X. Study on the relationship between the structure and activities of alkyl methacrylate–maleic anhydride polymers as cold flow improvers in diesel fuels. *Fuel Process. Technol.,* **2005**, *86*(6), 641-650.
[http://dx.doi.org/10.1016/j.fuproc.2004.05.011]

[40]   Aiyejina, A.; Chakrabarti, D.P.; Pilgrim, A.; Sastry, M.K.S. Wax formation in oil pipelines: A critical review. *Int. J. Multiph. Flow,* **2011**, *37*(7), 671-694.
[http://dx.doi.org/10.1016/j.ijmultiphaseflow.2011.02.007]

[41]   Jing, G.; Ye, P.; Zhang, Y. Pour point depressant for waxy crude oil: A mini-review. *Recent Innov. Chem. Eng.,* **2017**, *9*(2), 78-87.
[http://dx.doi.org/10.2174/2405520408666161021110008]

[42]   Subramanian, D.; Wu, K.; Firoozabadi, A. Ionic liquids as viscosity modifiers for heavy and extra-heavy crude oils. *Fuel,* **2015**, *143*, 519-526.
[http://dx.doi.org/10.1016/j.fuel.2014.11.051]

[43]   Castro, L.V.; Vázquez, F. Copolymers as flow improvers for Mexican crude oils. *Energy Fuels,* **2008**, *22*(6), 4006-4011.
[http://dx.doi.org/10.1021/ef800448a]

[44]   Castro, L.V.; Flores, E.A.; Vazquez, F. Terpolymers as Flow Improvers for Mexican Crude Oils. *Energy Fuels,* **2011**, *25*(2), 539-544.
[http://dx.doi.org/10.1021/ef101074m]

[45]   Fanchi, J.R., Ed. *Kokal. S. Petroleum Engineering Handbook: General Engineering*; Society of Petroleum Engineering: Texas, **2006**.

[46]   Kokal, S. Crude-oil Emulsions: A-State-of-the-Art Review. *SPE Production and Facililities.,* **2005**, 5-13.

[47]   Guerrero, S.; Parra, L.J.; Abreu, E.; Montefusco, L.; Gil, L. Orimulsion. *Interciencia,* **2004**, *29*, 180-181.

[48]   Miller, C.A.; Srivastava, R.K. The combustion of Orimulsion and its generation of air pollutants. *Pror. Energy Combust. Sci.,* **2000**, *26*(2), 131-160.
[http://dx.doi.org/10.1016/S0360-1285(99)00014-3]

[49]   Salager, J.L. Microemulsions. In: *Handbook of Detergents-Part. Surfactant Science Series 82*; Broze, A.G., Ed.; Marcel Dekker: New York, **1999**; pp. 253-302.

[50]   Salager, J.L.; Marquez, N.; Graciaa, A.; Lachaise, J. Partitioning of ethoxylated octylphenol surfactants in microemulsion-oil-water systems: influence of temperature and relation between partitioning coefficient and physicochemical formulation. *Langmuir,* **2000**, *16*(13), 5534-5539.
[http://dx.doi.org/10.1021/la9905517]

[51]   Martínez-Palou, R.; Reyes, J.; Cerón-Camacho, R.; Ramírez-de-Santiago, M.; Villanueva, D.; Vallejo, A.A.; Aburto, J. Study of the formation and breaking of extra-heavy-crude-oil-in-water emulsions—A proposed strategy for transporting extra heavy crude oils. *Chem. Eng. Process.,* **2015**, *98*, 112-122.
[http://dx.doi.org/10.1016/j.cep.2015.09.014]

[52]   Cerón-Camacho, R.; Aburto, J.A.; Montiel, L.E.; Martínez-Palou, R. Microwave-assisted organic synthesis *versus* conventional heating. A comparative study for Fisher glycosidation of monosaccharides. *C. R. Chim.,* **2013**, *16*(5), 427-432.
[http://dx.doi.org/10.1016/j.crci.2012.12.011]

[53]   Cerón-Camacho, R.; Aburto, J.; Montiel, L.E.; Flores, E.A.; Cuellar, F.; Martínez-Palou, R. Efficient Microwave-Assisted Synthesis of Ionic Esterified Amino Acids. *Molecules,* **2011**, *16*(10), 8733-8744.
[http://dx.doi.org/10.3390/molecules16108733]

[54]   Cerón-Camacho, R.; Martínez-Palou, R.; Chávez-Gómez, B.; Cuéllar, F.; Bernal-Huicochea, C.; Clavel, J.C.; Aburto, J. Synergistic effect of alkyl-O-glucoside and -cellobioside biosurfactants as

effective emulsifiers of crude oil in water. A proposal for the transport of heavy crude oil by pipeline. *Fuel,* **2013**, *110*, 310-317.
[http://dx.doi.org/10.1016/j.fuel.2012.11.023]

[55]   Martínez-Palou, R.; Cerón-Camacho, R.; Chávez, B.; Vallejo, A.A.; Villanueva-Negrete, D.; Castellanos, J.; Karamath, J.; Reyes, J.; Aburto, J. Demulsification of heavy crude oil-in-water emulsions: A comparative study between microwave and thermal heating. *Fuel,* **2013**, *113*(113), 407-414.
[http://dx.doi.org/10.1016/j.fuel.2013.05.094]

[56]   Vallejo-Cardona, A.A.; Martínez-Palou, R.; Chávez-Gómez, B.; García-Caloca, G.; Guerra-Camacho, J.; Cerón-Camacho, R.; Reyes-Ávila, J.; Karamath, J.R.; Aburto, J. Demulsification of crude oil-i--water emulsions by means of fungal spores. *PLoS One,* **2017**, *12*(2), e0170985.
[http://dx.doi.org/10.1371/journal.pone.0170985] [PMID: 28234917]

[57]   Flores, E.A.; Martínez-Palou, R.; Guzmán, D.; Likhanova, N.V. Ionic liquids as viscosity reducers of heavy crude oils. *Canadian Pat. Appl.,* (2) *708*, 416.

[58]   Martínez-Palou, R.; Flores Oropeza, E.A.; Guzmán-Lucero, D.; Lykhanova, N.V. Ionic liquids as viscosity reducers of heavy crude oils. Canadian Pat. Appl. (2), 708-416.

[59]   Reyes, J.; Martínez-Palou, R.; Cerón, R.; Vallejo, A.A.; Rodriguez, R.; Chávez, B.; Aburto, J. Process for demulsification of crude oil in water emulsions by means of natural or synthetic amino acid-based demulsifiers. *Canadian Pat.,* 2, 852, 865.

[60]   Martínez-Palou, R.; Cerón, R.; Vallejo, A.A.; Reyes, J.; Bernal, C.; Clave, J.; Ramirez de Santiago, M.; Aburto, J. Improved process for obtaining ionic amino acid esters. *Canadian Pat.,* 2, 903, 564.

[61]   Martínez-Palou, R.; Cerón, R.; Vallejo, A.A.; Reyes, J.; Bernal, C.; Ramirez de Santiago, M.; Aburto, J. Process for Demulsification of Crude Oil in Water Emulsions by means natural or synthetic aminoacid-based demulsifiers. U.S. Pat. 9,677,009 B2.

[62]   Hu, Y.F.; Guo, T.M. Effect of the structures of ionic liquids and alkylbenzene-derived amphiphiles on the inhibition of asphaltene precipitation from CO2-injected reservoir oils. *Langmuir,* **2005**, *21*(18), 8168-8174.
[http://dx.doi.org/10.1021/la050212f] [PMID: 16114918]

[63]   Hu, Y.F.; Guo, T.M. Effect of the structures of ionic liquids and alkylbenzene-derived amphiphiles on the inhibition of asphaltene precipitation from CO2-injected reservoir oils. *Langmuir,* **2005**, *21*(18), 8168-8174.
[http://dx.doi.org/10.1021/la050212f] [PMID: 16114918]

[64]   Ogunlaja, A.S.; Hosten, E.; Tshentu, Z.R. Dispersion of Asphaltenes in Petroleum with Ionic Liquids: Evaluation of Molecular Interactions in the Binary Mixture. *Ind. Eng. Chem. Res.,* **2014**, *53*(48), 18390-18401.
[http://dx.doi.org/10.1021/ie502672q]

[65]   Zheng, C.; Brunner, M.; Li, H.; Zhang, D.; Atkin, R. Dissolution and suspension of asphaltenes with ionic liquids. *Fuel,* **2019**, *238*, 129-138.
[http://dx.doi.org/10.1016/j.fuel.2018.10.070]

[66]   Bera, A.; Agarwal, J.; Shah, M.; Shah, S.; Vij, R.K. Recent advances in ionic liquids as alternative to surfactants/chemicals for application in upstream oil industry. *J. Ind. Eng. Chem.,* **2020**, *82*, 17-30.
[http://dx.doi.org/10.1016/j.jiec.2019.10.033]

[67]   Murillo-Hernández, J.A.; Aburto, J. Current Knowledge and Potential Applications of Ionic Liquids in the Petroleum Industry. In: *Ionic Liquids*; Applications and Perpectives, InTech, **2011**.
[http://dx.doi.org/10.5772/13974]

[68]   Murillo-Hernández, J.A.; García-Cruz, I.; López-Ramírez, S.; Durán-Valencia, C.; Domínguez, J.M.; Aburto, J. Aggregation Behavior of Heavy Crude Oil−Ionic Liquids Solutions by Fluorescence Spectroscopy. *Energy Fuels,* **2009**, *23*(9), 4584-4592.

[http://dx.doi.org/10.1021/ef9004175]

[69] Hernandez-Bravo, R.; Miranda, A.D.; Martínez-Mora, O.; Domínguez, Z.; Martínez-Magadan, J.M.; García-Chávez, R.; Domínguez-Esquivel, J.M. Calculation of the Solubility Parameters by COSMO-RS Methods and its Influences on Asphaltenes-Ionic Liquids Interactions. *Ind. Eng. Chem. Res.,* **2017**, *56*, 5107-5115.
[http://dx.doi.org/10.1021/acs.iecr.6b05035]

[70] Hernández-Bravo, R.; Miranda, A.D.; Martínez-Magadán, J.M.; Domínguez, J.M. Experimental and Theoretical Study on Supramolecular Ionic Liquid (IL)–Asphaltene Complex Interactions and Their Effects on the Flow Properties of Heavy Crude Oils. *J. Phys. Chem. B,* **2018**, *122*(15), 4325-4335.
[http://dx.doi.org/10.1021/acs.jpcb.8b01061] [PMID: 29587484]

[71] Hassanshahi, N.; Hu, G.; Li, J. Application of Ionic Liquids for Chemical Demulsification: A Review. *Molecules,* **2020**, *25*(21), 4915.
[http://dx.doi.org/10.3390/molecules25214915] [PMID: 33114253]

[72] Díaz Velázquez, H.; Guzmán-Lucero, D.; Martínez-Palou, R. Microwave-assisted demulsification in Petroleum Industry. A critical review. *J. Dispers. Sci. Technol.,* In press
[http://dx.doi.org/10.1080/01932691.2022.2049293]

CHAPTER 6

Application of ionic liquids as Corrosion Inhibitors in the Oil Industry

Abstract: The oil industry presents corrosion problems from crude oil extraction and transportation to the refining process, making it a highly demanding industry in terms of corrosion inhibitors. This chapter reviews the concepts related to the topic of corrosion and reviews the advances in the use of ILs to mitigate corrosion in different corrosive environments typical of this industry.

Keywords: Ionic liquids, Corrosion inhibitors, Acid environment, Basic environment, Carbon steel, Copper, Bottle test, Pipelines, Electrochemical techniques, Corrosion mechanism.

INTRODUCTION

Since most pieces of equipment, pipelines, mechanical parts, etc. are made of iron alloys, specialized personnel working at different production and processing centers, where these elements are omnipresent, know that they are unavoidably susceptible to become rusty due to diverse operation and environmental conditions. Because of this constant threat, industrial and commercial performance is seriously affected, which is reflected in economic and material drawbacks. As an example of the seriousness of this situation, the Gross Domestic Product [1] of countries such as the United States of America and Great Britain has been hit with corrosion-related losses that represent 3 or 4% of it.

Unfortunately, the negative effects caused by corrosion are not only represented by those mentioned above, but also by human affections, for the integrity and sanitary conditions of employed personnel and industrial-center inhabitants are at stake. In this sense, one of the economic activities that depict the complex picture formed by all the non-positive outcomes resulting from rust is the extraction, transportation, and refining of oil [2].

Among the solutions that have been envisaged to diminish the occurrence of corrosion in industrial environments, the synthesis of chemical compounds known as corrosion inhibitors (CIs) is found and their application has proven to be highly effective for this purpose. Although different inorganic compounds such as metal salts, nanomaterials and lanthanides [3] display corrosion inhibition properties, organic compounds have turned out to be highly effective at the industrial level and especially in the oil industry. Some of the most common families of organic compounds studied as CIs are amides with long alkyl chains (fatty amides) [4, 5], some pyridinic compounds [6 - 8], 1,3-azoles [9 - 11], different types of polymers [12], drugs [13], and imidazolines [14 - 16], the latter being the most common inhibitors used to prevent equipment and carbon steel or copper pipes that are in contact with oil from being corroded.

These compounds, when used in formulations in which they are added at concentrations between 50-500 ppm, can slow down considerably the corrosion process.

Table **6.1** shows characteristic structures of organic compounds that have been evaluated as CIs in the oil industry.

Table 6.1. Some families of chemical compounds employed as CIs in the Oil Industry.

Chemical Family	Chemical Structure	Applications	Refs.
Primary amines and diamines	Alkylamines (n =2-18) $CH_3\text{-}(CH_2)_n\text{-}NH_2$ Diamines (n = 2-8) $H_2N\text{-}(CH_2)_n\text{-}NH_2$	Mainly used as CIs of carbon steel and other metals in acidic media.	[17, 18]
	Cycloalkylamines (structure) H_2N		[19]
	Aromatic amines (X = H, NO_2, CH_3, Cl, COOH) X—(ring)—NH_2		[20]
Secondary, tertiary and ethoxylated amines	Bencylamines $HN\text{—}R$ (structure) Ethoxylated amines $CH_3\text{-}(CH_2)_n\text{-}NH\text{-}(OCH_2CH_2)_n$	Used as CIs of carbon steel in acidic media.	[21, 22]

(Table 6.1) cont.....

Chemical Family	Chemical Structure	Applications	Refs.
Oximes	Alkylic Arylic	Used as CIs of carbon steel in acidic media.	[23]
Nitriles	Alkylic Arylic	Used as CIs of carbon steel and other alloys in acidic media.	[24]
Ureas and thioureas	X= O, S, R = alkyl, aryl	Used as CIs of carbon steel and other alloys in acidic media.	[25]
Amides and thioamides	Amides Thioamides R, R' = alkyl	Mainly used as CIs of carbon steel in acidic media.	[26, 27]
Imidazoles	R = akyl, aryl	As CIs of carbon steel and copper alloys in basic media.	[28]
Benzoazoles	X = N-R, S, O	As CIs of carbon steel and copper alloys in basic media.	[29]
Imidazolines	R = akyl, aryl; X = NH$_2$, NHR, OH	CIs of carbon steel in CO$_2$ and other acid environments.	[30]

Chemical Family	Chemical Structure	Applications	Refs.
Pyridines	X = CH$_3$, Br, OR	Mainly used as CIs of carbon steel in acidic media.	[31]
Triazoles	R = akyl, aryl	As CIs of carbon steel and copper alloys in basic media.	[32]
Benzotriazoles	R = akyl, aryl	As CIs of copper alloys and steel in basic media and carbon steel in acid media.	[33]
Tetrazoles	R = akyl, aryl	As CIs of carbon steel and copper alloys in basic media.	[34]
Polyvinyls	R-(CH=CH)$_n$ R y R' = alkyl, aryl, heterocycles	As CIs of carbon steel in acidic media.	[35]
Polyethers	R-(OCH$_2$CH$_2$)n R = akyl, aryl	As CIs of carbon steel in acidic media.	[36]
Amino acids	H$_2$N—C=O, OH (R)	As CIs of carbon steel in acidic and basic media.	[37, 38]

Corrosion Theory

Corrosion can be described by the electrochemical reaction occurring between a metal and the medium surrounding it, causing gradual, uninterrupted, and permanent damage of the material physical and chemical features [39].

To have a better understanding of this phenomenon, it is necessary to know which the corrosion sources are. What follows is a classification and description of such sources:

Physical Corrosion

It is caused by blows on the material, mechanical effort or material weakening.

Chemical Corrosion

Elements such as oxygen, sulfur, fluorine, chlorine, or some gases that are in contact with metal pieces under specific operating and environmental conditions are fundamental in the occurrence of this type of corrosion.

Electrochemical Corrosion

Fundamental factors like the presence of an electrolyte and the electrical interaction between anodic and cathodic regions are required for this spontaneous phenomenon to take place.

Microbiological Corrosion

Bacteria and microalgae are responsible for this type of metal decay by living on the surface of metallic materials and employing nitrogen, oxygen, hydrogen and/or carbon from the surrounding medium to carry out their vital organic functions, producing waste matter that has corrosive properties. It is known that microorganisms can be active in different environments like drinking water, seawater, petroleum, and oil emulsions, which widens their corrosion capacity.

Corrosion can also be classified according to the medium in which the exposed materials are found; different forms of corrosion occur: uniform, pitting, erosion, stress corrosion, cavitation, galvanic and hydrogen embrittlement. Knowledge of the corrosion form helps understand the phenomenon to provide possible solutions to counteract the corrosive process.

Uniform or General Corrosion

This swift typical corrosion type extends itself practically homogeneously on metallic surfaces; in this process, the anodic site is where material loss takes place. At first sight, the surface damage looks uniform, but irregularities can be spotted (Fig. **6.1**) [40].

Pitting Corrosion

This kind of corrosion does not affect the whole surface but localized spots that suffer corrosion at high rate. These void spaces known as "pits" can extend themselves inside the material and are produced by corrosion solids or neutralization salts. Naturally, these surface points are prone to undergo corrosion (Fig. **6.2**) [40].

Fig. (6.1). General corrosion.

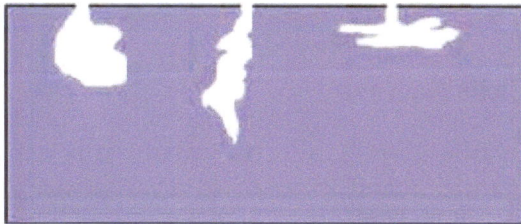

Fig. (6.2). Pitting corrosion.

Erosion Corrosion

When a metallic element is in contact with substances flowing at high rate, it is very likely that material erosion will occur. Consequences of this event are thickness loss and revealing of "fresh" metal surface that is susceptible of becoming rusty again and again through fast attack. Damage is worsened when solid elements part of liquid mixtures passing through pipelines are harder than the inner metal walls, thus boosting the erosion phenomenon (Fig. **6.3**) [41].

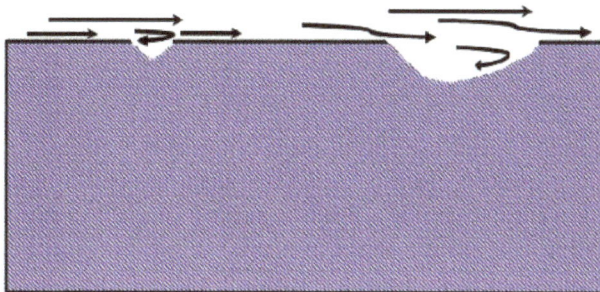

Fig. (6.3). Erosion corrosion.

Stress Corrosion

The submission of a metallic material to both forces that put to the test its mechanical resistance and highly acid or basic media can provoke the occurrence of cracks that can extend their "fingers" through and under the metal surface (Fig. **6.4**) [42].

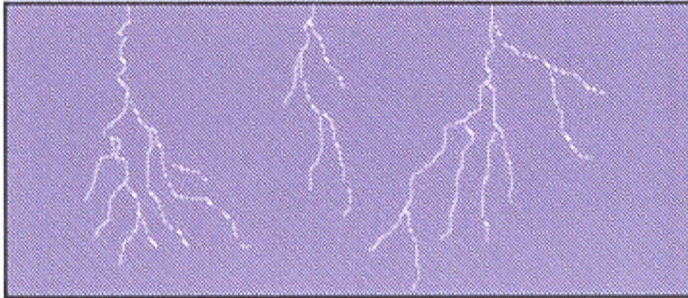

Fig. (6.4). Stress corrosion.

Galvanic or Bimetallic Corrosion

For galvanic or bimetallic corrosion to take place, it is necessary that a pair of metals of different nature, when exposed to an aggressive medium, present a potential difference that makes electrons move between both metallic pieces. The anode (active metal) role is played by the metal specimen with lower resistance and the cathode (noble metal) role is performed by the more resistant metal. This electron transfer is so damaging that it boosts corrosion of the reactive metal; the aggressiveness of galvanic corrosion stems from both the potential difference between metal samples and their implied surface area ratio (Fig. **6.5**) [43].

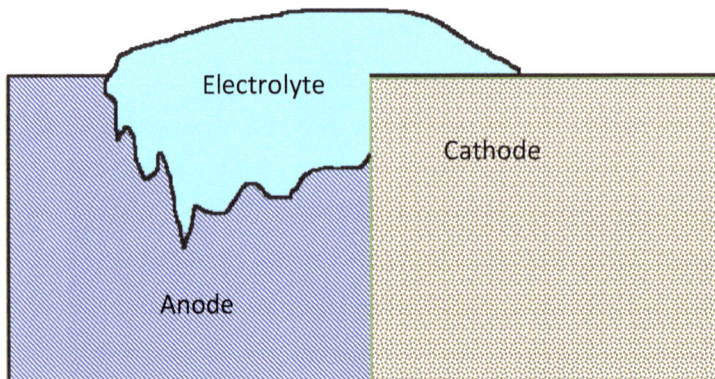

Fig. (6.5). Galvanic or bimetallic corrosion.

Cavitation Corrosion

When a metallic specimen is immersed in a liquid substance, vapor bubbles can gather on its surface. By bursting, such tiny "balloons" cause sequential "drumming" through the phenomenon known as cavitation that produces tiny but under-the-surface erosion (Fig. **6.6**) [44].

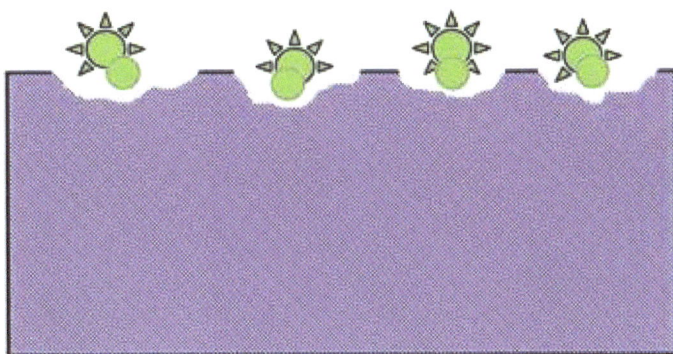

Fig. (6.6). Cavitation corrosion.

Corrosion *via* Embrittlement and Hydrogen Blistering

When hydrogen is generated on a metal surface in aqueous medium through reduction reactions, atomic hydrogen can reach under the material surface. The presence of metal defects will facilitate the production of molecular hydrogen, which by being enclosed inside the metal, can apply sufficient pressure to give way to blistering that prompts the appearance of microcracks. These defects take place most of the times in basic media having sulfides and/or cyanides, which is the case of catalytic plants performing refining processes [45].

The majority of refinery process complexes employ iron as reactive metal, which is the main element of steel, water as electrolyte and acids, salts, bases, and oxygen, among others, as oxidizing agents. To diminish corrosion taking place on the daily basis in the oil industry, CIs have become the most popular compounds put into practice, which in addition can be designed according to each process stage (Fig. **6.7**) [46].

Corrosion Monitoring in the Oil Industry

The corrosion phenomenon has to be followed through without interruption by the implementation of a program aimed at confirming that adequate protection is being achieved.

In order to have insight into the performance of an anticorrosion program, the protection effects have to be measured by means of quantification methods through which it is possible to spot factors to be improved and launch more effective inhibition methods.

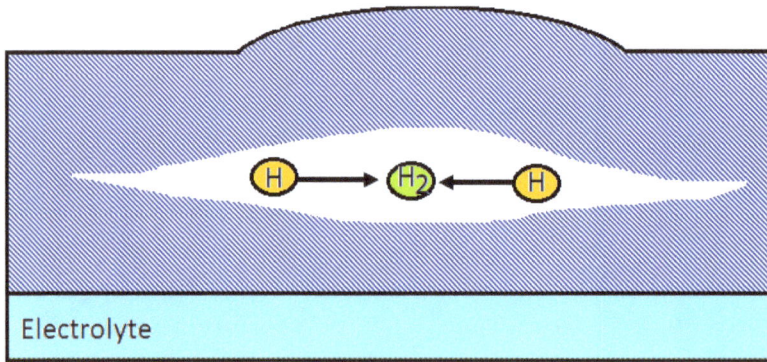

Fig. (6.7). Hydrogen blister corrosion.

What follows is a list of corrosion screening ways:

1. Load monitoring, which consists in analyzing the load entering the plant to know some of its characteristics and detect the content of possible corrosive agents.

2. Corrosive agent monitoring, which analyzes sour water from accumulators, considering pH, chlorides, sulfides, ammonium, thiocyanates and cyanides.

3. Corrosion monitoring can be done as follows:

a. By placing steel coupons where corrosion needs to be quantified by means of gravimetric tests.
b. By employing corrosimetric probes, which are set at the points to be studied and by means of a corrosimeter that is connected to the specimen, a quantity of current is sent and depending on the quantity of received current, it is known if there has been corrosion and at which rate.
c. By carrying out the analysis of Fe and Cu presence in accumulator sour water.
d. Using H probes located inside absorption towers, in the second step compression accumulator, gas-amine contact tower and in the amine section inlet liquid separator.

Corrosion Control

To partially reduce the corrosion problem, several preventive measures are taken, such as:

Cathodic protection. This is an effective method to control corrosion in buried structures or structures immersed in an electrolyte; according to the operation mode, they are classified as impressed current and sacrificial anodes [47].

Corrosion protection coating is mainly used to form a physical barrier between the corrosive environment and the structure to be protected. It is mainly used on metallic elements exposed to the atmosphere.

Corrosion Inhibitors

As previously discussed, these are chemical compounds or their formulations, which when added in low amounts (ppm) to the aggressive medium, effectively reduce the corrosion rate. CIs are mainly employed inside pipelines, containers and process equipment.

Action mechanism of Inhibitors

The action mechanisms of inhibitors are the following:

a. By adsorption, forming a film that adheres itself to the metal surface.
b. By inducing the formation of a corrosion product, such as iron sulfide, which is a passivating agent.
c. By changing the characteristics of the medium; it can be by producing protective precipitates and removing or deactivating a corrosive compound.

It is known that organic molecules prevent corrosion from happening by means of being adsorbed on a metal surface, thus creating a film separating the material and the surroundings. In this way, the polar part of the molecule is linked straight to the surface and the non-polar part moves vertically away from the material to be protected, repelling the action of aggressive molecules and preventing chemical and electrochemical damage.

Evaluation of the Toxicity of Inhibitors

Some of the factors to be considered in the synthesis of a CI are the toxicity and environment pollution potentials of the active agent and other formula constituents.

The European Economic Community appointed the Paris Commission (PARCOM) to provide guidance on environmental pollution control, ecosystem protection and toxicity assessment of starting matters, products and industry wastes.

PARCOM put into practice a standard environmental test that takes into account 3 points:

1. Toxic potential of the formula

For this factor, the Lethal Concentration 50 (LC50), which is the amount that kills 50% of evaluated organisms, or the EC50, which is the amount capable of exerting negative effects on living beings *e.g.* the concentration that reduces 50% of the emission capacity of luminescent bacteria or of the size or average mass of some microorganisms, can be used.

Toxicity can be sorted out based on the LC50 value considering the categories listed in Table **6.2** [1].

Table 6.2. Toxicity classification of a chemical compound according to its LC50.

Category	LC_{50}
Supertoxic	5 mg/kg weight or less
Extremely toxic	5-50 mg/kg
Highly toxic	50-500 mg/kg
Moderately toxic	0.5-5 g/kg
Slightly toxic	5-15 g/kg
Practically non-toxic	More than 15 g/kg

Toxicity tests of CIs can be carried out using a minimum of 3 different species and optimal time (Table **6.3**) [1].

Table 6.3. Standardized parameters for the development of toxicity tests.

Group	Preferred Species	Test
Algae	*Skeletonema costatum*	72 h EC_{50}
Fish and crustaceans	*Acartia tonsa*	48 h LC_{50}
Parasites	*Coropium voluntaros*	10 days LC_{50}

2. Biodegradation

The formula constituents have to be evaluated. How much time the chemical compounds CI consists need to undergo degradation is established through the biodegradation capacity. To this end, the standardized Early-life Stage Toxicity marine test (OECD) has to be put into practice. The accepted top value is above 60% after 28 days.

3. Bioaccumulation

The purpose of this test is to establish how much product can be stored by a given organism, which is done by the partition coefficient, for it is associated with the cell/water interface (Eq.(1)):

$P_{o/w}$ = Octanol concentrationWater concentration (1)

If $P_{O/W}$ is high, the possibility of a given chemical substance to penetrate the cell membrane and proceed to biological accumulation is also high.

Techniques for Evaluating the Performance of Corrosion inhibitors

In general, the techniques used to evaluate the performance of CIs are divided into gravimetric and electrochemical techniques [48 - 50].

Gravimetric Techniques

These techniques are based on the weight loss that a material undergoes when subjected to a corrosive medium. At industrial level, corrosimetric cores are used, which are located at strategic points in the pipelines and ducts of refineries, as previously mentioned. At laboratory level, the most standardized tests are known as the wheel and loss-in-weight rotating cylinder methods. Both methods consist in weighing a wheel or a cylinder of the material whose corrosion rate is to be determined and placing it in a highly saline solution that promotes corrosion of the material. Then, the material begins to rotate for a certain time and at the end, the initial minus the final weight difference is determined.

Electrochemical Techniques

Polarization Resistance

This technique is used to measure the absolute corrosion rate. This measurement can be performed in less than ten minutes, but it can only measure general corrosion and does not measure localized corrosion.

Tafel Extrapolation Cursive

This method considers the use of potentiodynamic curves and is employed for measuring the current at which corrosion occurs (how fast) and identifying what type of kinetics controls the electrochemical reaction rate can be established. Tafel extrapolation requires a single electrochemical process whose rate-determining step is the charge transfer (activation control). It works with experimental (steady-state) and apparent polarization curves, and since they do not exactly match the corrosion process, it is necessary to perturb the equilibrium system to record them on passive surfaces or materials; this method makes it possible to determine parameters such as the passivation and repassivation potentials and the potential range of the passive zone to determine the efficiency with which a passive film protects a metal from corrosion and to characterize the behavior of the passive film.

Linear polarization techniques and Tafel diagrams obtained through these studies are the most widely used electrochemical techniques because they are simple and fast, generally used to carry out corrosion analysis and determine the corrosion rate and type of corrosion (Fig. **6.8**).

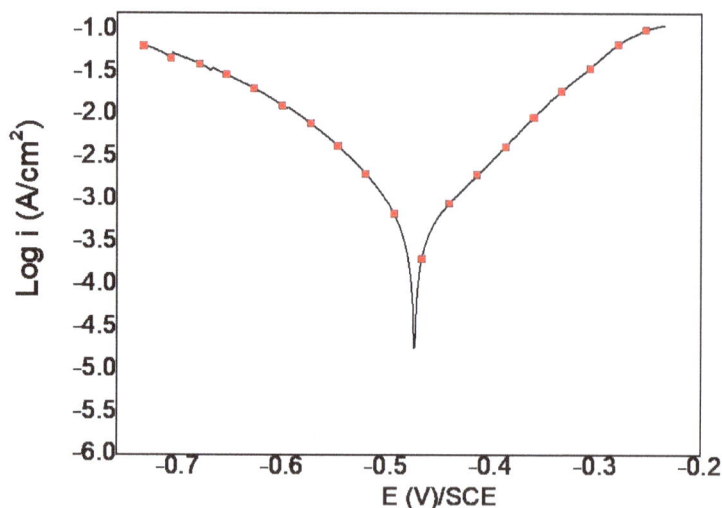

Fig. (6.8). Example of a Tafel polarization plot.

Electrochemical Impedance Spectroscopy

This is a method used in corrosion studies and is based on the use of an alternating current signal applied to an electrode to determine the corresponding response and allows the behavior of the metal-solution interface to be deduced, offering a complete view of the corrosive phenomena taking place. The major advantage of this technique is the possibility of measuring independent corrosion processes occurring simultaneously on the same material as a function of time.

The most common way of presenting the information obtained from the Electrochemical impedance spectroscopy is through Nyquist plot, where the negative imaginary Impedance-Z" is plotted *versus* the real part of the impedance Z' (Fig. **6.9**).

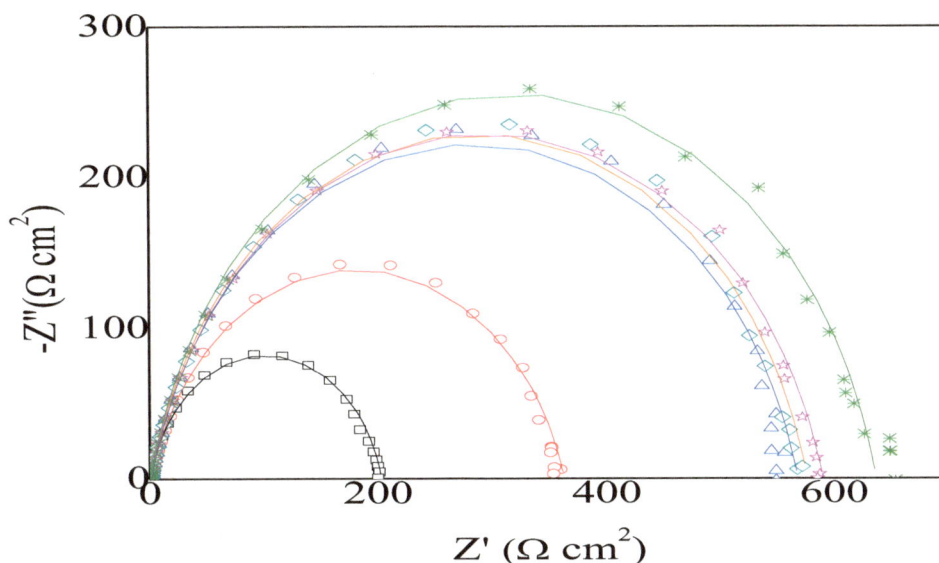

Fig. (6.9). Example of a Nyquist diagram.

Electrochemical Noise

This technique is used to determine localized corrosion (*e.g.* pitting) by monitoring the open circuit electrical potential and electrical current density as a function of time. It is necessary to have an experienced handling of this technique to carry out an adequate analysis and interpretation of results, while electrochemical impedance spectroscopy is a rather complex technique that requires a comprehensive mathematical analysis; additionally, the equipment and experimental design are sophisticated. Taking into consideration these qualities, it is the technique that gives us more information about the system under study, since parameters such as thickness of the corrosion layer can be obtained. Electrochemical noise enables us to determine the type of corrosion that is occurring.

Research in the Development of CIS for the Oil Industry

The development of CIs for different fields of application is a very active area of research [51]. In case of the Oil Industry, there has been a great deal of research on inhibitors for copper [52], and carbon steel [53], which are the most used metals in this industry.

The Mexican Petroleum Institute (IMP) has been the pioneer institution in Mexico that has dedicated resources for more than 30 years to innovation, research, and development of CIs for the Oil Industry. As a result of this work, a series of innovations and technological developments have been patented in the area of products for chemical treatments in refineries (demulsifiers, detergents, gasoline additives, etc.) and particularly CI formulations have been developed, many of which are successfully applied at industrial level.

Among the most recent results, Marín et al. have patented nationally and internationally [54 - 56] a CI composition that has shown great effectiveness in oil pipelines, with results not only at laboratory level, but also through industrial scale evaluations in Petróleos Mexicanos (PEMEX, Mexican Petroleum Company) facilities in which the high effectiveness of this CI formulation was endorsed.

The IMP has published several scientific articles about the synthesis and evaluation of chemical compounds as CIs [57 - 65].

Ionic Liquids as Corrosion Inhibitors

ILs have attracted great interest in recent years due to their outstanding performance as CIs [66, 67].

The IMP has been engaged in the search for low-toxicity chemical compounds as an alternative for the development of environmentally friendly CIs, *e.g.* in the article titled Evaluation of Corrosion Inhibitors Synthesized from Fatty Acids and Fatty Alcohols isolated from Sugar Cane Wax by R. Martínez-Palou, J. Rivera, L. G. Zepeda, M. A. Hernandez, J. Marín, and A. Estrada (Corrosion 2004, 60, 465-470), the synthesis and use of chemical products extracted from sugar cane, such as waxes and waxy alcohols as efficient alternatives to control corrosion using natural products are discussed [68].

In this sense, in the book chapter entitled Environmentally Friendly Corrosion Inhibitors (Chapter 19) by R. Martínez-Palou, O. Olivares, and N. V. Likhanova (Progress in Corrosion Inhibitors. Intech, 2014, p. 431-465), different examples of chemical compounds with the potential to be used as efficient low-toxicity CIs are discussed; in addition to natural products, examples of different works demonstrating the effectiveness of some ILs for their application are also shown. In this context, the IMP has published some works; for example, in 2010, Likhanova and collaborators evaluated a series of ILs containing imidazolium and pyridinium cations as CIs in acidic environments for light steel, which are found typically in some areas of the Oil Industry.

In particular, 1,3-dioctadecylimidazolium bromide and N-octadecylpyridinium bromide were studied, which were synthesized by conventional and microwave-assisted reactions, respectively. The ILs tested as CIs showed corrosion protection efficiency within 82-88% at 100 ppm for mild steel in 1 M aqueous solution of sulfuric acid. The standard free energy indicated that corrosion inhibition occurred by a chemical adsorption process. The surface analysis (SEM, EDX) completed by XRD and Mössbauer spectroscopy indicated the presence of carbon species, which allowed us to suggest an inhibition mechanism.

Corrosion inhibition was within the range of 82-88% at 100 ppm for mild steel in aqueous sulfuric acid solution, which increased with inhibitor concentration in the range of 10-100 ppm. 1,3-bis(octadecyl) imidazolium bromide (ImDC18Br) provided a better inhibition effect than octadecyl pyridinium bromide (PyC18Br). These chemicals modified the anodic and cathodic reactions, which made them to be considered as mixed-type CIs. The Langmuir isotherm described the chemical sorption undergone by these CIs on the mild steel material [69].

In 2011, Guzman-Lucero and co-workers synthesized by conventional synthesis and employing microwave irradiation five imidazolium cation-type ILs with unsaturated N1 nitrogen bound chains and bromide as anion (IL1IL5). The compounds were tested in water/H_2SO_4 (1 M) as CIs of carbon steel. The weight loss and polarization curves indicated that the inhibition efficiency increased with concentration and was also significantly dependent on the size of the alkyl chain attached to N3. The relatively high inhibitory properties (88-95%) exhibited by IL4 at 25-40°C were attributed to a chemisorption process involving the following adsorption of protonated imidazolium molecules on the anodic and cathodic sites. Surface analysis indicated a considerable reduction in the production of corrosion products after adding IL4. The inhibition mechanism of these surfactants reveals the production of a protecting film on the metal surface, which is depicted in Fig. **(6.10)** [70].

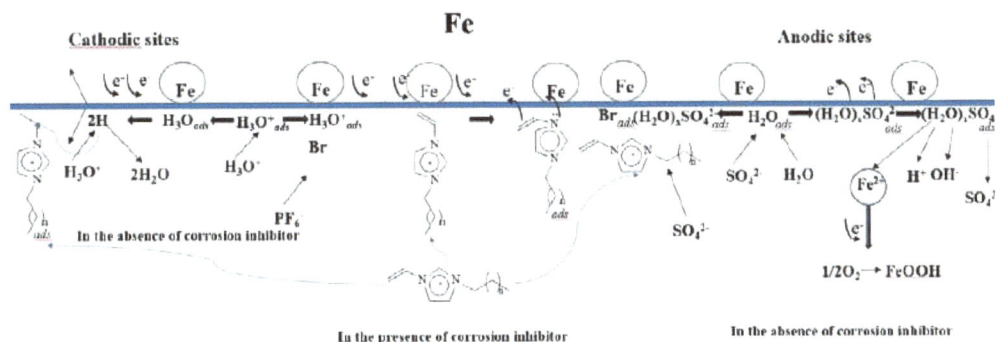

Fig. (6.10). Schematic representation of the corrosion inhibition mechanism by vinylimidazolium ILs. [Adapted with permission from Guzmán-Lucero, D.; Olivares-Xometl, O.; Martínez-Palou, R.; Likhanova, N. V.; Domínguez-Aguilar, M. A.; Garibay-Febles, V. Synthesis of Selected Vinylimidazolium Ionic Liquids and Their Effectiveness as Corrosion Inhibitors for Carbon Steel in Aqueous Sulfuric Acid. *Ind. Eng. Chem. Res.* **2011**, *50*, 7129–7140. Copyright 2022 American Chemical Society].

In another work by this research group, three new ILs were evaluated as CIs: 1,2-dimethyl-3-decylimidazolium iodide (DDI), N-triethylmethylammonium acetate (TMA) and N-triethylmethylammonium laurate (TML). Two acid corrosive media, sulfuric acid (1.0 M) and hydrochloric acid (1.0 M) solutions were used to inhibit the corrosion of API 5L X-52 steel. CIs were evaluated by Tafel polarization and polarization resistance (PR). The Tafel polarization results confirmed that DDI, TMA and TML inhibited the corrosion of steel and functioned as mixed inhibitors in both corrosive media. In 1.0 M H_2SO_4, DDI showed the best inhibition efficiency in 1.0 M H_2SO_4, reaching 95% while TMA in 1.0 M HCl showed the best inhibitory efficiency, reaching a maximum of 70% [71].

In another work, two ILs, 1,3-dibenzylimidazolium acetate and 1,3-dibenzylimidazolium dodecanoate (DBImL), were evaluated as CIs in 1.0 M HCl and H_2SO_4 solutions. DBImL was tested in the same corrosive media described in the previous work. The performed electrochemical studies showed that, as in previous cases, these materials were physisorbed, preventing the attack of the ions in the solution. Both inhibitors displayed mixed inhibitor action by inhibiting both the anodic metal dissolution and cathodic reaction [72].

Ammonium-based ILs with amphiphilic character due to the presence of long alkyl chains have also been evaluated by IMP researchers. In 2018, Arellanes-Lozada et al. synthesized and evaluated the ILs methyltrioctylammonium methyl sulfate and trimethyltetradecylammonium methyl sulfate (TTA) as CIs of API-X52 steel in HCl (1 M).

The corrosion inhibition capacity was assessed by means of electrochemical methods and it was elicited that methyltrioctylammonium methyl sulfate and TTA worked as mixed CIs by preventing corrosion according to temperature, immersion time and IL amount. It was identified that 40 °C was the best thermal condition, which was associated with the good stability of methyl sulfate negative ions.

The adsorption process of the CIs on the steel surface implied intense electrostatic exchanges including the metal, chloride ions, methyl sulfate anions and quaternary ammonium cations in the ILs.

At 40 °C, the inhibitors played an outstanding protection role due to the conformation of negatively charged methyl sulfate whose negligible nucleophilic features allowed better thermal resistance. The positively charged ions were attracted toward the metal surface by means of the (R)4-N + radical and the lengthy alkyl strings positioned themselves on the medium side, which promoted the formation of an additional, positively charged film located at opposing position. This configuration reorganization constrained the displacement of H^+ and Cl^- in the direction of the metal, which prevented API-X52 from being corroded.

These researchers paid special attention to ammonium cation ILs, because, among other reasons, these compounds allow relatively cheaper ILs to be generated than those derived from heterocyclic cations. In 2018, two ILs were synthesized and evaluated as CIs of API-X52 steel in HCl (1 M): methyltrioctylammonium methyl sulfate and TTA [73].

In another recent work, two new ILs, *N*-ethyl-*N,N,N*-trihexylammonium adipate (CPA6) and N-ethyl-*N,N,N*-trioctylammonium ethyl sulfate (ESA8), were synthesized and the corrosion inhibition effect of both compounds on API-X60 steel in H_2SO_4 (1 M) was evaluated by gravimetric and electrochemical techniques (Tafel polarization curves and polarization resistance). The experimental results showed that the ILs were adsorbed on the API-X60 steel and blocked the surface-active sites, which considerably reduced the corrosion rate in relation to the control material subjected to the corrosive medium without the protection of the evaluated chemicals. The ILs behaved as mixed-type corrosion inhibitors and their adsorption process obeyed the Langmuir isotherm model. The adsorption phenomenon was studied by XPS, DRIFT and SEM/EDS techniques. Quantum chemical parameters of the studied ILs were obtained, calculated in vacuum and water systems, and correlated with experimental results. The effect of organic anions on ILs was analyzed and it was concluded that the ethyl sulfate anion presented better corrosion inhibition properties than the adipate anion; thus, the ESA8 compound performed better as CI than CPA6. The performance increased with increasing concentration. The calculated standard Gibbs free energy described an adsorption process by electrostatic interactions on the steel surface. A SEM micrograph confirmed a significant difference in surface damage when protected by the inhibitors compared to when unprotected (Fig. **6.11**) [74].

Through Rp and Tafel analysis, it was found that corrosion products were refrained from approaching the metal surface thanks to the structural impairment created by IL molecules adsorbed on active spots. The protection action occurred by rejecting H_2O and corrosive species from the metallic contact area, which augmented θ, which in turn depends on the inhibitor molecular weight.

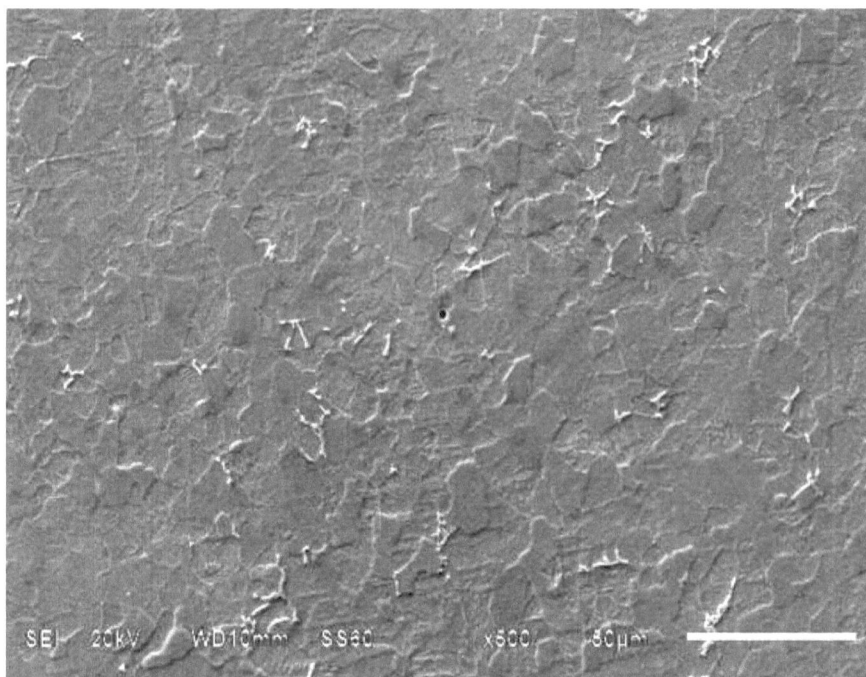

Fig. (6.11). SEM micrograph image obtained for this work.

The βa and βc values and E_{corr} revealed that [$N_{1,2,2,2}$][C_n, n = 6, 9 and 12] chemicals worked as mixed inhibitors with positively-charged-end affinity. In this case, it was found that the extent of the alkyl part of the carboxylate anion (HOOC-C_nH_{2n}-COO−) affected negligibly the η of the [$N_{1,2,2,2}$][C_9 and C_{12}] chemicals according to the inhibitor amount. Nevertheless, [$N_{1,2,2,2}$][C_6] highlighted the consequence of the alkyl part extent in the negatively charged carboxylate ion present in the quaternary-ammonium-derived cations. By means of theory-based calculations, it was found that [$N_{1,2,2,2}$][C_n, n = 6, 9 and 12] interacted with the metal through coordinated chemical linkages featuring -COO− (placed at the negatively-charged ion) and nitrogen (placed at the positively-charged ion). Furthermore, hydrophilic activity in the corrosion environment was promoted remarkably by the alkyl chain. The presence of N atoms was revealed by XPS, which came from the IL positively-charged ions. As for the C1s spectrum, the energy values were related to CC/C-H and C/N linkages. In turn, DRIFTS helped identify the C\\O signal from negatively-charged ions (Fig. **6.12**) [75].

Fig. (6.12). Structures of cation and anions of the dicarboxylic ILs employed as inhibitors. [Adapted with permission from Olivares-Xometl, O.; Lijanova, I. V.; Likhanova, N. V.; Arellanes-Lozada, P.; Hernández-Cocoletzi, H.; J. Arriola-Morales, J. Theoretical and experimental study of the anion carboxylate in quaternary-ammonium-derived ionic liquids for inhibiting the corrosion of API X60 steel in 1 M H_2SO_4. *J. Mol. Liq.* **2020**, *318*, 114075.Copyright 2022 Elsevier and Copyright Clearance Center].

In this work, the authors proposed a mechanism for the possible interaction between the dicarboxylic anion and the metal surface that could explain the performance shown by these compounds (Fig. **6.13**) [75].

Fig. (6.13). Mechanism of the possible interaction between the dicarboxylic anion and the metal surface. [Adapted with permission from Olivares-Xometl, O.; Lijanova, I. V.; Likhanova, N. V.; Arellanes-Lozada, P.; Hernández-Cocoletzi, H.; J. Arriola-Morales, J. Theoretical and experimental study of the anion carboxylate in quaternary-ammonium-derived ionic liquids for inhibiting the corrosion of API X60 steel in 1 M H_2SO_4. *J. Mol. Liq.* **2020**, *318*, 114075.Copyright 2022 Elsevier and Copyright Clearance Center].

In another work by the same research group, the synthesis and characterization of four new ILs derived from ammonium, which were evaluated as CIs of API 5L X60 steel exposed to sulfuric acid (1.0 M), were investigated. The featured ILs had not been previously reported as CIs, and in order to determine their η, the potentiodynamic polarization and weight loss techniques were used whereas the characterization of the steel surface was performed by SEM/EDX, AFM and FTIR [76].

In 2019, three other ILs with ammonium cation were synthesized and evaluated as CIs; in this case, triethylmethyl ammonium laurate (TAL), triethylmethyl ammonium anthranilate (TAA) and triethylmethyl ammonium oleate (TAO). The evaluation was carried out using API 5L X52 steel in H_2SO_4 (0.5 M) using the inhibitors within a concentration interval ranging from 10 to 100 ppm. The electrochemical evaluations reported an inhibition efficiency of 78% for TAL, 73% for TAA and 63% for TAO [77].

Arellanes-Lozada et al. synthesized two imidazolium-type ILs containing iodide as anion and tested them as CIs of API 5L X52 steel in H_2SO_4 (1 M). The ILs 1,2-dimethyl-3-butylimidazolium iodide ([DBIM]$^+$I$^-$) and 1,2-dimethyl-3-propylimidazolium iodide ([DPIM]$^+$I$^-$) showed corrosion inhibition properties that were enhanced with increasing concentration and length of the alkyl chain of the ILs.

The surface analyses by SEM, AFM and computed tomography showed that in the presence of [DBIM]$^+$I$^-$ and [DPIM]$^+$I$^-$, the corrosion rate was slower by the adsorption effect of IL molecules on the steel surface [78].

This year, Likhanova et al. synthesized and evaluated two poly(ionic liquids) (PolyILs) derived from alkylimidazolium imidazolate as CIs of API 5L X52 steel exposed to oil reservoir production water. The corrosion inhibition performance of the PolyILs was investigated by potentiodynamic polarization (PDP), electrochemical impedance spectroscopy, scanning electron microscopy (SEM), atomic force microscopy (AFM), and X-ray photoelectron spectroscopy (XPS). According to the PDP analysis, it was concluded that the PolyILs could be classified as mixed-type CIs with anodic tendency, featuring maximal IE of 80%. The adsorption mechanism occurred through the blocking of active sites by means of electrostatic attraction forces and formation of chemical bonds with the metal of the surface. The SEM and AFM surface analyses evidenced that the metal surface in the presence of the PolyILs was less damaged by steel corrosion. Finally, the XPS studies confirmed that the CI interaction with the steel surface reduced the amount of produced corrosion products such as FeOH and/or $FeCO_3$ [79].

CONCLUDING REMARKS

As we have seen in the present chapter, oil refineries present different corrosive environments for which corrosion inhibitors have been developed, and different families of chemical compounds, including natural compounds friendly to the environment are capable of considerably slowing down the corrosion process when added in low concentrations.

ILs have shown that they can have an important performance as CIs in diverse aggressive media present in the oil industry. Due to their properties, these ILs can protect metal surfaces of equipment and pipelines having a low toxicity. ILs are often mixed-type inhibitors, which are adsorbed to metal surfaces through physical or chemical adsorption. Imidazolium derivatives with alkylic chains greater than 10 carbon atoms show good anticorrosive activity, especially in the medium of sulfuric acid. Recently, scientific interest has revolved around corrosion inhibitors based on ILs with anion based on halide-free carboxylic acids. We expect that in the near future, ionic liquids will gain an important place among the chemicals of wide industrial application to prevent corrosion in the oil industry.

REFERENCES

[1]   Quraishi, M.A.; Chauhan, D.S.; Ansari, F.A. Development of environmentally benign corrosion inhibitors for organic acid environments for oil-gas industry. *J. Mol. Liq.,* **2021**, *329*, 115514.
      [http://dx.doi.org/10.1016/j.molliq.2021.115514]

[2]   Sastri, V.S. *Corrosion Inhibitors Principles and Applications*; John Wiley & Sons: New York, **1998**.

[3]   Bethencourt, M.; Botana, F.J.; Calvino, J.J.; Marcos, M.; Rodríguez-Chacón, M.A. Lanthanide compounds as environmentally-friendly corrosion inhibitors of aluminium alloys: a review. *Corros. Sci.,* **1998**, *40*(11), 1803-1819.
      [http://dx.doi.org/10.1016/S0010-938X(98)00077-8]

[4]   Olivares, O.; Likhanova, N.V.; Gómez, B.; Navarrete, J.; Llanos-Serrano, M.E.; Arce, E.; Hallen, J.M. Electrochemical and XPS studies of decylamides of α-amino acids adsorption on carbon steel in acidic environment. *Appl. Surf. Sci.,* **2006**, *252*(8), 2894-2909.
      [http://dx.doi.org/10.1016/j.apsusc.2005.04.040]

[5]   Elsharif, A.M.; Abubshait, S.A.; Abdulazeez, I.; Abubshait, H.A. Synthesis of a new class of corrosion inhibitors derived from natural fatty acid: 13☐Docosenoic acid amide derivatives for oil and gas industry. *Arab. J. Chem.,* **2020**, *13*(5), 5363-5376.
      [http://dx.doi.org/10.1016/j.arabjc.2020.03.015]

[6]   Abd El-Maksoud, S.A.; Fouda, A.S. Some pyridine derivatives as corrosion inhibitors for carbon steel in acidic medium. *Mater. Chem. Phys.,* **2005**, *93*(1), 84-90.
      [http://dx.doi.org/10.1016/j.matchemphys.2005.02.020]

[7]   Ergun, Ü.; Yüzer, D.; Emregül, K.C. The inhibitory effect of bis-2,6-(3,5-dimethylpyrazolyl)pyridine on the corrosion behaviour of mild steel in HCl solution. *Mater. Chem. Phys.,* **2008**, *109*(2-3), 492-499.
      [http://dx.doi.org/10.1016/j.matchemphys.2007.12.023]

[8]   Noor, E.A. Evaluation of inhibitive action of some quaternary N-heterocyclic compounds on the

corrosion of Al–Cu alloy in hydrochloric acid. *Mater. Chem. Phys.,* **2009**, *114*(2-3), 533-541.
[http://dx.doi.org/10.1016/j.matchemphys.2008.09.065]

[9]   Likhanova, N.V.; Martínez-Palou, R.; Veloz, M.A.; Matías, D.J.; Reyes-Cruz, V.E.; Höpfl, H.; Olivares, O. Microwave-assisted synthesis of 2-(2-pyridyl)azoles. Study of their corrosion inhibiting properties. *J. Heterocycl. Chem.,* **2007**, *44*(1), 145-153.
[http://dx.doi.org/10.1002/jhet.5570440123]

[10]   Popova, A.; Christov, M.; Zwetanova, A. Effect of the molecular structure on the inhibitor properties of azoles on mild steel corrosion in 1M hydrochloric acid. *Corros. Sci.,* **2007**, *49*(5), 2131-2143.
[http://dx.doi.org/10.1016/j.corsci.2006.10.021]

[11]   Antonijević, M.M.; Milić, S.M.; Petrović, M.B. Films formed on copper surface in chloride media in the presence of azoles. *Corros. Sci.,* **2009**, *51*(6), 1228-1237.
[http://dx.doi.org/10.1016/j.corsci.2009.03.026]

[12]   Tallman, D.E.; Spinks, G.; Dominis, A.; Wallace, G.G. Electroactive conducting polymers for corrosion control. *J. Solid State Electrochem.,* **2002**, *6*(2), 73-84.
[http://dx.doi.org/10.1007/s100080100212]

[13]   Gece, G. Drugs: A review of promising novel corrosion inhibitors. *Corros. Sci.,* **2011**, *53*(12), 3873-3898.
[http://dx.doi.org/10.1016/j.corsci.2011.08.006]

[14]   Shamsa, A.; Barmatov, E.; Hughes, T.L.; Hua, Y.; Neville, A.; Barker, R. Hydrolysis of imidazoline based corrosion inhibitor and effects on inhibition performance of X65 steel in CO2 saturated brine. *J. Petrol. Sci. Eng.,* **2022**, *208*, 109235.
[http://dx.doi.org/10.1016/j.petrol.2021.109235]

[15]   Olivares-Xometl, O.; Likhanova, N.V.; Martínez-Palou, R.; Domínguez-Aguilar, M.A. Electrochemistry and XPS study of an imidazoline as corrosion inhibitor of mild steel in an acidic environment. *Mater. Corros.,* **2009**, *60*(1), 14-21.
[http://dx.doi.org/10.1002/maco.200805044]

[16]   Liu, F.G.; Du, M.; Zhang, J.; Qiu, M. Electrochemical behavior of Q235 steel in saltwater saturated with carbon dioxide based on new imidazoline derivative inhibitor. *Corros. Sci.,* **2009**, *51*(1), 102-109.
[http://dx.doi.org/10.1016/j.corsci.2008.09.036]

[17]   Kumar, H.; Kumari, M. Experimental and Theoretical investigation of 3,3′-diamino dipropyl amine: Highly efficient corrosion inhibitor for carbon steel in 2 N HCl at normal and elevated temperatures. *J. Mol. Struct.,* **2020**, 129598.
[http://dx.doi.org/10.1016/j.molstruc.2020.129598]

[18]   Manivel, A.; Ramkumar, S.; Wu, J.J.; Asiri, A.M.; Anandan, S. Exploration of (S)-4,5,6,- -tetrahydrobenzo[d]thiazole-2,6-diamine as feasible corrosion inhibitor for mild steel in acidic media. *J. Environ. Chem. Eng.,* **2014**, *2*(1), 463-470.
[http://dx.doi.org/10.1016/j.jece.2014.01.018]

[19]   Kumar, H.; Dhanda, T. Cyclohexylamine an effective corrosion inhibitor for mild steel in 0.1 N H_2SO_4: Experimental and theoretical (molecular dynamics simulation and FMO) study. *J. Mol. Liq.,* **2020**, 114847.
[http://dx.doi.org/10.1016/j.molliq.2020.114847]

[20]   Mousaa, I.M. Gamma irradiation processed (epoxidized soybean fatty acids/ρ-substituted aromatic amines) adducts as corrosion inhibitors for UV–curable steel coatings. *Prog. Org. Coat.,* **2017**, *111*, 220-230.
[http://dx.doi.org/10.1016/j.porgcoat.2017.06.009]

[21]   Gao, G.; Liang, C.H.; Wang, H. Synthesis of tertiary amines and their inhibitive performance on carbon steel corrosion. *Corros. Sci.,* **2007**, *49*(4), 1833-1846.
[http://dx.doi.org/10.1016/j.corsci.2006.08.014]

[22]   El Basiony, N.M. AmrElgendy, El-Tabey, A. E.; Al-Sabagh, A. M.; El-Hafez, G. M. A.; El-raouf, M. A.; Migahed, M. A. Synthesis, characterization, experimental and theoretical calculations (DFT and MC) of ethoxylated aminothiazole as inhibitor for X65 steel corrosion in highly aggressive acidic media. *J. Mol. Liq.,* **2019**, 111940.
       [http://dx.doi.org/10.1016/j.molliq.2019.111940]

[23]   Li, X.; Deng, S.; Xie, X. Experimental and theoretical study on corrosion inhibition of oxime compounds for aluminium in HCl solution. *Corros. Sci.,* **2014**, *81*, 162-175.
       [http://dx.doi.org/10.1016/j.corsci.2013.12.021]

[24]   Fitoz, A.; Nazır, H. Özgür (nee Yakut), M.; Emregül, E.; Emregül, K. C. An experimental and theoretical approach towards understanding the inhibitive behavior of a nitrile substituted coumarin compound as an effective acidic media inhibitor. *Corros. Sci.,* **2018**, *133*, 451-464.
       [http://dx.doi.org/10.1016/j.corsci.2017.10.004]

[25]   Deng, S.; Li, X.; Xie, X. Hydroxymethyl urea and 1,3-bis(hydroxymethyl) urea as corrosion inhibitors for steel in HCl solution. *Corros. Sci.,* **2014**, *80*, 276-289.
       [http://dx.doi.org/10.1016/j.corsci.2013.11.041]

[26]   Özcan, M.; Dehri, İ. Electrochemical and quantum chemical studies of some sulphur-containing organic compounds as inhibitors for the acid corrosion of mild steel. *Prog. Org. Coat.,* **2004**, *51*(3), 181-187.
       [http://dx.doi.org/10.1016/j.porgcoat.2004.07.017]

[27]   Hosseini, S.M.A.; Salari, M.; Jamalizadeh, E.; Khezripoor, S.; Seifi, M. Inhibition of mild steel corrosion in sulfuric acid by some newly synthesized organic compounds. *Mater. Chem. Phys.,* **2010**, *119*(1-2), 100-105.
       [http://dx.doi.org/10.1016/j.matchemphys.2009.08.029]

[28]   Mishra, A.; Aslam, J.; Verma, C.; Quraishi, M.A.; Ebenso, E.E. Imidazoles as highly effective heterocyclic corrosion inhibitors for metals and alloys in aqueous electrolytes: A review. *J. Taiwan Inst. Chem. Eng.,* **2020**, *114*, 341-358.
       [http://dx.doi.org/10.1016/j.jtice.2020.08.034]

[29]   Vernack, E.; Costa, D.; Tingaut, P.; Marcus, P. DFT studies of 2-mercaptobenzothiazole and 2-mercaptobenzimidazole as corrosion inhibitors for copper. *Corr. Sci.,* **2020**, *174*, 108840.
       [http://dx.doi.org/10.1016/j.corsci.2020.108840]

[30]   Shamsa, A.; Barker, R.; Hua, Y.; Barmatov, E.; Hughes, T.L.; Neville, A. Performance evaluation of an imidazoline corrosion inhibitor in a CO2-saturated environment with emphasis on localised corrosion. *Corros. Sci.,* **2020**, *176*, 108916.
       [http://dx.doi.org/10.1016/j.corsci.2020.108916]

[31]   Verma, C.; Rhee, K.Y.; Quraishi, M.A.; Ebenso, E.E. Pyridine based N-heterocyclic compounds as aqueous phase corrosion inhibitors: A review. *J. Taiwan Inst. Chem. Eng.,* **2020**, *117*, 265-277.
       [http://dx.doi.org/10.1016/j.jtice.2020.12.011]

[32]   Fernandes, C.M.; Alvarez, L.X.; dos Santos, N.E.; Maldonado Barrios, A.C.; Ponzio, E.A. Green synthesis of 1-benzyl-4-phenyl-1H-1,2,3-triazole, its application as corrosion inhibitor for mild steel in acidic medium and new approach of classical electrochemical analyses. *Corros. Sci.,* **2019**, *149*, 185-194.
       [http://dx.doi.org/10.1016/j.corsci.2019.01.019]

[33]   Onyeachu, I.B.; Solomon, M.M. Benzotriazole derivative as an effective corrosion inhibitor for low carbon steel in 1 M HCl and 1 M HCl + 3.5 wt% NaCl solutions. *J. Mol. Liq.,* **2020**, *313*, 113536.
       [http://dx.doi.org/10.1016/j.molliq.2020.113536]

[34]   El-Askalany, A.H.; Mostafa, S.I.; Shalabi, K.; Eid, A.M.; Shaaban, S. Novel tetrazole-based symmetrical diselenides as corrosion inhibitors for N80 carbon steel in 1 M HCl solutions: Experimental and theoretical studies. *J. Mol. Liq.,* **2016**, *223*, 497-508.
       [http://dx.doi.org/10.1016/j.molliq.2016.08.088]

[35]   Gu, T.; Liu, X.; Chai, W.; Li, B.; Sun, H. A preliminary research on polyvinyl alcohol hydrogel: A slowly-released anti-corrosion and scale inhibitor. *J. Petrol. Sci. Eng.,* **2014**, *122*, 453-457.
       [http://dx.doi.org/10.1016/j.petrol.2014.08.005]

[36]   Liu, G.; Xue, M.; Yang, H. Polyether copolymer as an environmentally friendly scale and corrosion inhibitor in seawater. *Desalination,* **2017**, *419*, 133-140.
       [http://dx.doi.org/10.1016/j.desal.2017.06.017]

[37]   Zhang, Q.H.; Hou, B.S.; Li, Y.Y.; Lei, Y.; Wang, X.; Liu, H.F.; Zhang, G.A. Two amino acid derivatives as high efficient green inhibitors for the corrosion of carbon steel in CO2-saturated formation water. *Corros. Sci.,* **2021**, *189*, 109596.
       [http://dx.doi.org/10.1016/j.corsci.2021.109596]

[38]   Aslam, R.; Mobin, M.; Huda, ; Obot, I.B.; Alamri, A.H. Ionic liquids derived from α-amino acid ester salts as potent green corrosion inhibitors for mild steel in 1M HCl. *J. Mol. Liq.,* **2020**, *318*, 113982.
       [http://dx.doi.org/10.1016/j.molliq.2020.113982]

[39]   The Multimedia Corrosion Guide [CD-ROM]. INSA, Lyon; **2013**.

[40]   Marcus, P.; Maurice, V.; Strehblow, H.H. Localized corrosion (pitting): A model of passivity breakdown including the role of the oxide layer nanostructure. *Corros. Sci.,* **2008**, *50*(9), 2698-2704.
       [http://dx.doi.org/10.1016/j.corsci.2008.06.047]

[41]   Levy, A.V. The erosion-corrosion behavior of protective coatings. *Surf. Coat. Tech.,* **1988**, *36*(1-2), 387-406.
       [http://dx.doi.org/10.1016/0257-8972(88)90168-5]

[42]   Sieradzki, K.; Newman, R.C. Stress-corrosion cracking. *J. Phys. Chem. Solids,* **1987**, *48*(11), 1101-1113.
       [http://dx.doi.org/10.1016/0022-3697(87)90120-X]

[43]   Song, G.; Johannesson, B.; Hapugoda, S.; StJohn, D. Galvanic corrosion of magnesium alloy AZ91D in contact with an aluminium alloy, steel and zinc. *Corros. Sci.,* **2004**, *46*(4), 955-977.
       [http://dx.doi.org/10.1016/S0010-938X(03)00190-2]

[44]   Al-Hashem, A.; Riad, W. The role of microstructure of nickel–aluminium–bronze alloy on its cavitation corrosion behavior in natural seawater. *Mater. Charact.,* **2002**, *48*(1), 37-41.
       [http://dx.doi.org/10.1016/S1044-5803(02)00196-1]

[45]   González, J.L.; Ramirez, R.; Hallen, J.M.; Guzmán, R.A. Hydrogen-Induced Crack Growth Rate in Steel Plates Exposed to Sour Environments. *Corrosion,* **1997**, *53*(12), 935-943.
       [http://dx.doi.org/10.5006/1.3290278]

[46]   Revie, W.; Uhlig, H.H. *Corrosion and corrosion control: An introduction to corrosion science and engineering*; Wiley-Interscience: New York, **2008**.
       [http://dx.doi.org/10.1002/9780470277270]

[47]   Sastri, V.S. *Green Corrosion Inhibitors. Theory and Practice*; John Wiley & Sons: Hoboken, NJ, **1998**.

[48]   Berradja, A. Electrochemical techniques for corrosion and tribocorrosion monitoring: Methods for the assessment of corrosion rates.*Corrosion Inhibitors*; IntechOpen, **2019**.
       [http://dx.doi.org/10.5772/intechopen.86743]

[49]   Ropital, F. Environmental degradation in hydrocarbon fuel processing plant: issues and mitigation.*Advances in Clean Hydrocarbon Fuel Processing*; Woodhead Publishing, **2011**, pp. 437-462.
       [http://dx.doi.org/10.1533/9780857093783.5.437]

[50]   Roy, D. Electrochemical techniques and their applications for chemical mechanical planarization (CMP) of metal films. In: *Advances in Chemical Mechanical Planarization (CMP)*; Woodhead Publishing, **2016**; pp. 47-89.

[51] Inhibitors, C. Principles and Recent Applications., **2017**.

[52] Fateh, A.; Aliofkhazraei, M.; Rezvanian, A.R. Review of corrosive environments for copper and its corrosion inhibitors. *Arab. J. Chem.,* **2020**, *13*(1), 481-544.
[http://dx.doi.org/10.1016/j.arabjc.2017.05.021]

[53] Finšgar, M.; Jackson, J. Application of corrosion inhibitors for steels in acidic media for the oil and gas industry: A review. *Corros. Sci.,* **2014**, *86*, 17-41.
[http://dx.doi.org/10.1016/j.corsci.2014.04.044]

[54] Marín-Cruz, J.; Vega-Paz, A.; Montiel-Sánchez, L. E.; Castillo-Cervantes, S.; Martínez Palou, R.; Estrada-Martínez, A.; Quej-Aké, L. M, Benitez Aguilar, J. L. R., Sánchez García V. Corrosion Inhibition Composition for Pipelines, Process of Elaboration and Synthesis. U.S. Pat. 10,676,829.

[55] Marín-Cruz, J.; Vega-Paz, A.; Montiel-Sánchez, L.E. Marín-Cruz, J., Vega-Paz, A., Montiel-Sánchez, L. E. Castillo-Cervantes, S.; Martínez Palou, R.; Estrada-Martínez, A.; Quej-Aké, L. M, Benitez Aguilar, J. L. R., Sánchez García V. Corrosion Inhibition Composition for Pipelines, Process of Elaboration and Synthesis. Appl. Pat. Germany 10 , **2014**. 222 031.9.

[56] Marín-Cruz, J.; Vega-Paz, A.; Montiel-Sánchez, L. E.; Castillo-Cervantes, S.; Martínez Palou, R.; Estrada-Martínez, A.; Quej-Aké, L. M.; Benitez-Aguilar, J. L. R.; Sánchez García V. Corrosion Inhibition Composition for Pipelines, Process of Elaboration and Synthesis. Appl. Pat. Canadian 2,880,361.

[57] Gómez, B.; Likhanova, N.V.; Domínguez-Aguilar, M.A.; Martínez-Palou, R.; Vela, A.; Gázquez, J.L. Quantum chemical study of the inhibitive properties of 2-pyridyl-azoles. *J. Phys. Chem. B,* **2006**, *110*(18), 8928-8934.
[http://dx.doi.org/10.1021/jp057143y] [PMID: 16671697]

[58] Likhanova, N.V.; Martínez-Palou, R.; Veloz, M.A.; Matías, D.J.; Reyes-Cruz, V.E.; Höpfl, H.; Olivares, O. Microwave-assisted synthesis of 2-(2-pyridyl)azoles. Study of their corrosion inhibiting properties. *J. Heterocycl. Chem.,* **2007**, *44*(1), 145-153.
[http://dx.doi.org/10.1002/jhet.5570440123]

[59] Fragoza-Mar, L.; Olivares-Xometl, O.; Domínguez-Aguilar, M.A.; Flores, E.A.; Arellanes-Lozada, P.; Jiménez-Cruz, F. Corrosion inhibitor activity of 1,3-diketone malonates for mild steel in aqueous hydrochloric acid solution. *Corros. Sci.,* **2012**, *61*, 171-184.
[http://dx.doi.org/10.1016/j.corsci.2012.04.031]

[60] Cruz, J.; Martínez, R.; Genesca, J.; García-Ochoa, E. Experimental and theoretical study of 1-(--ethylamino)-2-methylimidazoline as an inhibitor of carbon steel corrosion in acid media. *J. Electroanal. Chem. (Lausanne),* **2004**, *566*(1), 111-121.
[http://dx.doi.org/10.1016/j.jelechem.2003.11.018]

[61] Pérez-Navarrete, J.B.; Olivares-Xometl, C.O.; Likhanova, N.V. Adsorption and corrosion inhibition of amphiphilic compounds on steel pipeline grade API 5L X52 in sulphuric acid 1 M. *J. Appl. Electrochem.,* **2010**, *40*(9), 1605-1617.
[http://dx.doi.org/10.1007/s10800-010-0146-2]

[62] Olivares-Xometl, O.; Likhanova, N.V.; Domínguez-Aguilar, M.A.; Arce, E.; Dorantes, H.; Arellanes-Lozada, P. Synthesis and corrosion inhibition of α-amino acids alkylamides for mild steel in acidic environment. *Mater. Chem. Phys.,* **2008**, *110*(2-3), 344-351.
[http://dx.doi.org/10.1016/j.matchemphys.2008.02.010]

[63] Ramírez-Pérez, J.F.; Zamudio-Rivera, L.S.; Servín-Nájera, A.G.; Soto-Castruita, E.; Cerón-Camacho, R.; Cisneros-Dévora, R.; Oviedo-Roa, R.; Martínez-Magadán, J.M. Quantum modeling design of imidazoline-based corrosion inhibitors for oil industry applications. *Mater. Today Commun.,* **2021**, *27*, 102466.
[http://dx.doi.org/10.1016/j.mtcomm.2021.102466]

[64] Olivares-Xometl, O.; Likhanova, N.V.; Domínguez-Aguilar, M.A.; Hallen, J.M.; Zamudio, L.S.; Arce,

E. Surface analysis of inhibitor films formed by imidazolines and amides on mild steel in an acidic environment. *Appl. Surf. Sci.,* **2006**, *252*(6), 2139-2152.
[http://dx.doi.org/10.1016/j.apsusc.2005.03.178]

[65]   Olivares-Xometl, O.; Likhanova, N.V.; Domínguez-Aguilar, M.A.; Arce, E.; Dorantes, H.; Arellanes-Lozada, P. Synthesis and corrosion inhibition of α-amino acids alkylamides for mild steel in acidic environment. *Mater. Chem. Phys.,* **2008**, *110*(2-3), 344-351.
[http://dx.doi.org/10.1016/j.matchemphys.2008.02.010]

[66]   Ardakani, E.K.; Kowsari, E.; Ehsani, A.; Ramakrishna, S. Performance of all ionic liquids as the eco-friendly and sustainable compounds in inhibiting corrosion in various media: A comprehensive review. *Microchem. J.,* **2021**, *165*, 106049.
[http://dx.doi.org/10.1016/j.microc.2021.106049]

[67]   Popoola, L.T. Progress on pharmaceutical drugs, plant extracts and ionic liquids as corrosion inhibitors. *Heliyon,* **2019**, *5*(2), e01143.
[http://dx.doi.org/10.1016/j.heliyon.2019.e01143] [PMID: 30766932]

[68]   Martínez-Palou, R.; Rivera, J.; Zepeda, L.G.; Rodríguez, A.N.; Hernández, M.A.; Marín-Cruz, J.; Estrada, A. Evaluation of Corrosion Inhibitors Synthesized from Fatty Acids and Fatty Alcohols Isolated from Sugar Cane Wax. *Corrosion,* **2004**, *60*(5), 465-470.
[http://dx.doi.org/10.5006/1.3299242]

[69]   Likhanova, N.V.; Domínguez-Aguilar, M.A.; Olivares-Xometl, O.; Nava-Entzana, N.; Arce, E.; Dorantes, H. The effect of ionic liquids with imidazolium and pyridinium cations on the corrosion inhibition of mild steel in acidic environment. *Corros. Sci.,* **2010**, *52*(6), 2088-2097.
[http://dx.doi.org/10.1016/j.corsci.2010.02.030]

[70]   Guzmán-Lucero, D.; Olivares-Xometl, O.; Martínez-Palou, R.; Likhanova, N.V.; Domínguez-Aguilar, M.A.; Garibay-Febles, V. Synthesis of Selected Vinylimidazolium Ionic Liquids and Their Effectiveness as Corrosion Inhibitors for Carbon Steel in Aqueous Sulfuric Acid. *Ind. Eng. Chem. Res.,* **2011**, *50*(12), 7129-7140.
[http://dx.doi.org/10.1021/ie1024744]

[71]   Olivares-Xometl, O.; López-Aguilar, C.; Herrastí-González, P.; Likhanova, N.V.; Lijanova, I.; Martínez-Palou, R.; Rivera-Márquez, J.A. Adsorption and Corrosion Inhibition Performance by Three New Ionic Liquids on API 5L X52 Steel Surface in Acid Media. *Ind. Eng. Chem. Res.,* **2014**, *53*(23), 9534-9543.
[http://dx.doi.org/10.1021/ie4035847]

[72]   Lozano, I.; Mazario, E.; Olivares-Xometl, C.O.; Likhanova, N.V.; Herrasti, P. Corrosion behaviour of API 5LX52 steel in HCl and H2SO4 media in the presence of 1,3-dibencilimidazolio acetate and 1,3-dibencilimidazolio dodecanoate ionic liquids as inhibitors. *Mater. Chem. Phys.,* **2014**, *147*(1-2), 191-197.
[http://dx.doi.org/10.1016/j.matchemphys.2014.04.029]

[73]   Arellanes-Lozada, P.; Olivares-Xometl, O.; Likhanova, N.V.; Lijanova, I.V.; Vargas-García, J.R.; Hernández-Ramírez, R.E. Adsorption and performance of ammonium-based ionic liquids as corrosion inhibitors of steel. *J. Mol. Liq.,* **2018**, *265*, 151-163.
[http://dx.doi.org/10.1016/j.molliq.2018.04.153]

[74]   Likhanova, N.V.; Arellanes-Lozada, P.; Olivares-Xometl, O.; Hernández-Cocoletzi, H.; Lijanova, I.V.; Arriola-Morales, J.; Castellanos-Aguila, J.E. Effect of organic anions on ionic liquids as corrosion inhibitors of steel in sulfuric acid solution. *J. Mol. Liq.,* **2019**, *279*, 267-278.
[http://dx.doi.org/10.1016/j.molliq.2019.01.126]

[75]   Olivares-Xometl, O.; Lijanova, I.V.; Likhanova, N.V.; Arellanes-Lozada, P.; Hernández-Cocoletzi, H.; Arriola-Morales, J. Theoretical and experimental study of the anion carboxylate in quaternary-ammonium-derived ionic liquids for inhibiting the corrosion of API X60 steel in 1 M H_2SO_4. *J. Mol. Liq.,* **2020**, *318*, 114075.
[http://dx.doi.org/10.1016/j.molliq.2020.114075]

[76] Olivares-Xometl, O.; Álvarez-Álvarez, E.; Likhanova, N.V.; Lijanova, I.V.; Hernández-Ramírez, R.E.; Arellanes-Lozada, P.; Varela-Caselis, J.L. Synthesis and corrosion inhibition mechanism of ammonium-based ionic liquids on API 5L X60 steel in sulfuric acid solution. *J. Adhes. Sci. Technol.,* **2018**, *32*(10), 1092-1113.
[http://dx.doi.org/10.1080/01694243.2017.1397422]

[77] Likhanova, N.V.; Arellanes-Lozada, P.; Olivares-Xometl, O.; Lijanova, I.V.; Arriola-Morales, J.; Mendoza-Hérnandez, J.C.; Corro, G. Ionic Liquids with Carboxylic-Acid-Derived Anions Evaluated as Corrosion Inhibitors under Dynamic Conditions. *Int. J. Electrochem. Sci.,* **2019**, *14*, 2655-2671.
[http://dx.doi.org/10.20964/2019.03.71]

[78] Arellanes-Lozada, P.; Díaz-Jiménez, V.; Hernández-Cocoletzi, H.; Nava, N.; Olivares-Xometl, O.; Likhanova, N.V. Corrosion inhibition properties of iodide ionic liquids for API 5L X52 steel in acid medium. *Corros. Sci.,* **2020**, *175*, 108888.
[http://dx.doi.org/10.1016/j.corsci.2020.108888]

[79] Likhanova, N.V.; López-Prados, N.; Guzmán-Lucero, D.; Olivares-Xometl, O.; Lijanova, I.V.; Arellanes-Lozada, P.; Arriola-Morales, J. Some polymeric imidazolates from alkylimidazolium as corrosion inhibitors of API 5L X52 steel in production water. *J. Adhes. Sci. Technol.,* **2022**, *36*(8), 845-874.
[http://dx.doi.org/10.1080/01694243.2021.1939600]

CHAPTER 7

Ionic Liquids as Inhibitors of Hydrate Formation in Deepwater Wells

Abstract: The extraction of crude oil in deep waters represents a major technological challenge. One of the most common problems in these oil and gas extraction conditions is the formation of gas hydrates that can cause plugging and hinder the transportation of crude oil. For this reason, great efforts have been made in the development of hydrate inhibitors, among which ILs have shown to be a very promising alternative. In this chapter, the development and evaluation of ILs as hydrate inhibitors are discussed.

Keywords: Anti-agglomerant, Clusters, Deepwater, Imidazolium, Ionic liquids, Hydrate inhibitors, Long alkyl chain, Low temperature, Low toxicity, Methane, Oil, Pyridinium.

INTRODUCTION

Gas hydrates are chemical compounds formed and stabilized when a small molecule (such as methane, ethane, H_2S, CO_2 and others) is trapped inside a solid structure formed by water molecules that could agglomerate to form ice-like structures [1]. The formation of this type of hydrate is very common in deepwater oil wells, where low temperatures and pressures favor the formation of this type of cluster that generates many problems in the transport of hydrocarbons in these areas, since they plug the pipelines generating operational and safety consequences and production stoppages.

Structure of Methane Hydrates

By their structure, gas hydrates can be classified into three types [2]:

SI structure: It is a cubic structure of 12 Å, made up of 46 water molecules in which two types of "cages" stand out: pentagonal dodecahedral (512) and tetradecahedral (51262). They are most frequently found in nature and store only small gas molecules. (CH_4, C_2H_6, CO_2 y H_2S).

SII structure: It is also a cubic structure of 17.3 Å, with 136 water molecules and is made up of two types of cages: one of type 512 and one hexadecahedral

(51264). The most common molecules it can store are nitrogen, propane, and isobutane. This structure is typically found in natural gas production.

Structure H: Less common than the previous ones, it consists of three types of cages: a 512 type, an irregular dodecahedron (435663) and an irregular icosahedron (51268). Fig. (**7.1**) shows the structure of the hydrates described above.

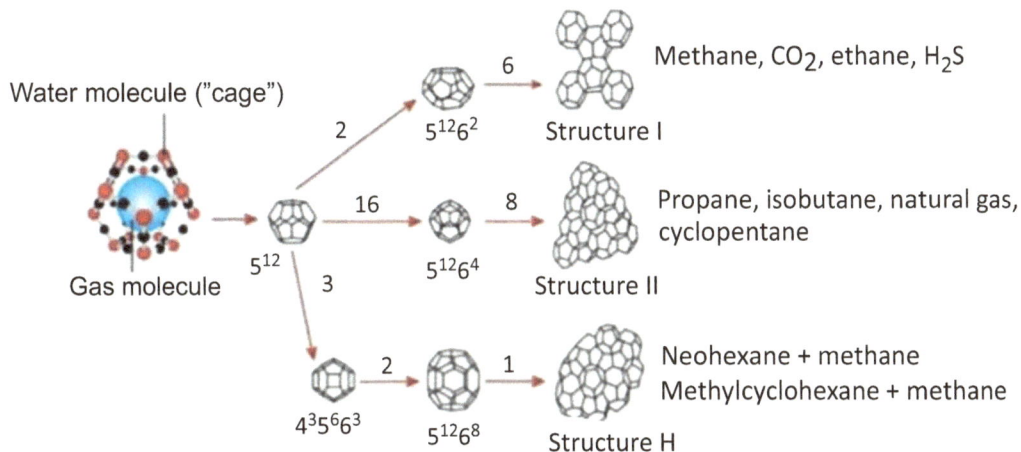

Fig. (**7.1**). Types of gas hydrates and their composition and structure. [Available *via* license: Creative Commons Attribution 4.0 International (https://creativecommons.org/licenses/by/4.0/)].

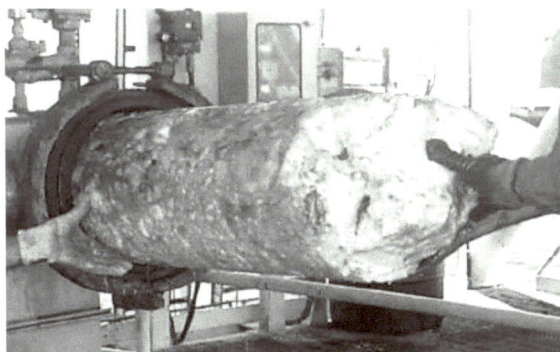

Fig. (**7.2**). Pipeline plugged by gas hydrates. [Available *via* license: Creative Commons Attribution 4.0 International (https://creativecommons.org/licenses/by/4.0/)].

The three types of hydrates have a structure similar to ice and when proliferating in production pipelines, they cause their plugging, significantly limiting and preventing the flow of hydrocarbons, which generates production stoppages. These blockages are favored when low temperatures (below 5 °C), high pressures, and the presence of high-velocity gases in the presence of water occur together. In the production and transport of natural gas in deep waters, this occurs at the seabed level, where these conditions are favored (Fig. **7.2**).

Different methods have been studied and applied to prevent hydrocarbon clathrate hydrate plugs, including maintaining a suitable temperature in pipelines by insulation, electrical heating, pressure reduction or water removal, but these methods are often very expensive [3].

Thermodynamic Equilibrium of Hydrate Formation

Hydrate formation is a crystallization process that takes place in two main stages: nucleation and hydrate crystal growth. Nucleation is a microscopic phenomenon and at this stage, the system reaches a metastable state in which the crystals (nuclei) agglomerate until they reach a critical size, whereupon the crystal growth stage begins [4].

A typical Pressure-Temperature thermodynamic diagram of hydrate-forming systems consists of: the hydrate envelope (HLV phase equilibrium curve), the hydrate stability region (left) and hydrate-free region (right), and the degree of system subcooling (ΔTsub). The ΔTsub is defined as the difference between the equilibrium temperature (Teq), and a temperature (Ti) in the hydrate stability region reached by the system and at which crystal growth occurs, at a given pressure (ΔTsub= Teq - Ti) (Fig. **7.3**) [5].

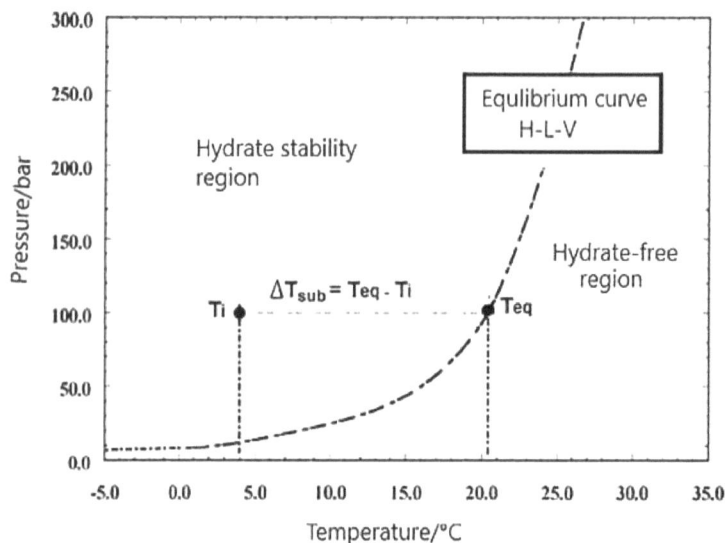

Fig. (7.3). Pressure *versus* temperature diagram typical of gas hydrate formation.

Chemicals to Prevent Gas Hydrate Formation

To control the formation and agglomeration of hydrocarbon clathrate hydrates, the oil industry has been using the addition of different chemicals such as thermodynamic hydrate inhibitors (THI, *e.g.* methanol and glycols) [3]. Because of the high concentration used, these chemicals represent an environmental risk and are not economically attractive because they require chemical regeneration facilities. Current research is focused on the development of new low-dosage products and, more recently, of low-dosage hydrate inhibitors known as kinetic hydrate inhibitors or Low Dosage Hydrate Inhibitors (LDHIs) and anti-agglomerant agents [6, 7].

LDHIs are pure compounds that can be mixed with others, showing a synergistic effect, and used at low concentrations (generally less than 3.0% per mass) to delay nucleation and, consequently, the formation of gas hydrates for periods of time that depend mainly on the hydrates [8, 9].

The most common compounds used as LDHIs are water-soluble polymeric molecules containing hydrophobic and hydrophilic functionalities, with high hydrate inhibition efficiency and high chemical stability [10]. The most common LDHIs are polymers based on *N*-Vinylcaprolactam (VCap), *N*-Vinylpyrrolidone and *N*-isopropylmethacrylamide monomers that are part of the most commonly employed commercial formulations to prevent gas hydrate formation in the present century and have generated a large number of patents [11 - 19]. Hyperbranched poly(ester amides) (HPEAs) based on diisopropanolamine and several carboxylic anhydrides have also been used [20].

Poly(ionic liquid)s (poly(IL)s) are polymerized ILs in which one or more ILs are repeated throughout the polymer chain. They may also contain units of ILs combined with units of other nonionic monomers. These compounds combine the advantages of ILs and in these ionic polymeric units, certain properties characteristic of ILs such as virtually zero vapor pressure, high thermal and chemical stability, high ionic conductivity, wide electrochemical window, catalytic properties and modulation of acid-base and solubility properties are preserved [21 - 24], while combining these properties with those characteristic of polymers in terms of processability (membrane formation, cross-linking, porous materials), durability and regenerability.

In recent years, Poly(IL)s have demonstrated extensive efficiency as environmental-friendly chemicals with different applications in the oil industry [8, 25]. Recently, the IMP developed novel copolymers [26], and ionic terpolymers obtained from VCap with polymerizable ionic monomers that have shown excellent performance as LDHIs [27]. For example, in 2017, in the work titled

Evaluation of copolymers from 1-vinyl-3-alkylimidazolium bromide and M VCap as inhibitors of clathrate hydrate formation (J. Nat. Gas Sci. Eng. 2017, 40, 114-125) by Rebolledo-Libreros, M. E.; Reza, J.; Trejo, A.; and Guzmán-Lucero, D. J [26], several copolymers of VCap combined with alkylimidazolium units, featuring linear and branched chains, were synthesized and evaluated, finding that these compounds displayed higher cloud points than the poly(VCap) homopolymer, which is the base of widely used commercial LDHIs. In this work, three experimental methods were employed to evaluate the hydrate inhibition performance of these ionic copolymers: the ability to inhibit the hydrate formation of tetrahydrofuran (THF) at atmospheric pressure and -0.5 °C; the ability to inhibit natural gas hydrate formation at high pressure (6 MPa and 1 °C); and the cloud point of the polymers in aqueous systems with varying salinity and the results were compared to the VCap homopolymer that is a recognized commercial LDHI.

All four evaluated copolymers were efficient in inhibiting the formation or agglomeration of both THF and natural gas hydrates with type II structure. The inclusion of an ionic functional group in the composition of the copolymers increased their hydrophilicity, favoring their solubility in water. The copolymer shown in Fig. (**7.4**) was completely soluble at concentration = 0.5 g/100, within a temperature interval ranging from 2 to 80 °C and its performance was very similar to that of the commercial inhibitor PVCap.

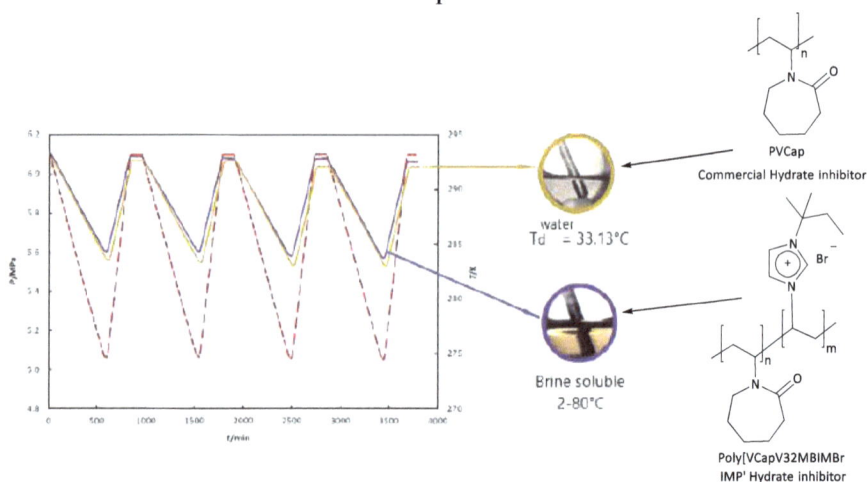

Fig. (7.4). Pressure and temperature variation as a function of time for type II natural gas hydrate in the presence of the copolymer poly[VCapV32MBIMBr] and PVCap (polymer concentration: 0.5 g/100 g water).

On the other hand, in the paper Inhibition of structure II hydrate formation by salt-tolerant N-vinyl lactam-based terpolymers (J. Nat. Gas Sci. Eng. 2018, 56,

175–192) by Reza, J.; Trejo, A.; Rebolledo-Libreros, M. E.; and Guzmán-Lucero, D [27], all IMP researchers, five VCap terpolymers, hydroxyethylmethacrylates or acrylamide derivatives and alkylimidazolium bromides or benzylammonium chloride were synthesized, which like the previously described homopolymers, also showed higher cloud points than a poly (VCap) homopolymer widely recognized as a hydrate inhibitor. Likewise, in this work, three experimental methods were employed to evaluate the hydrate inhibition performance of these ionic copolymers: the ability to inhibit hydrate formation of (THF at atmospheric pressure and -0.5 °C; the ability to inhibit natural gas hydrate formation at high pressure (6 MPa and 1 °C and subcooling at 16 °C for the natural gas sample study); and determination of the cloud point of the polymers in aqueous systems with varying salinity (at atmospheric pressure and temperature between 2 and 80 °C) and the results were compared with those of the VCap homopolymer that is a recognized commercial LDHI. The performance of these compounds as hydrate inhibitors and anti-caking agents shows that these compounds have a wide potential for industrial application. Fig. (7.5) shows, as an example, a T *vs* P curve for one of the two terpolymers formed by a blend of PVCap, vinylbenzyltrimethyl- ammonium and acrylamide (poly[VCapVBTMABrAA]), which presented better performance than the PVCap homopolymer preventing the formation of structure II hydrates.

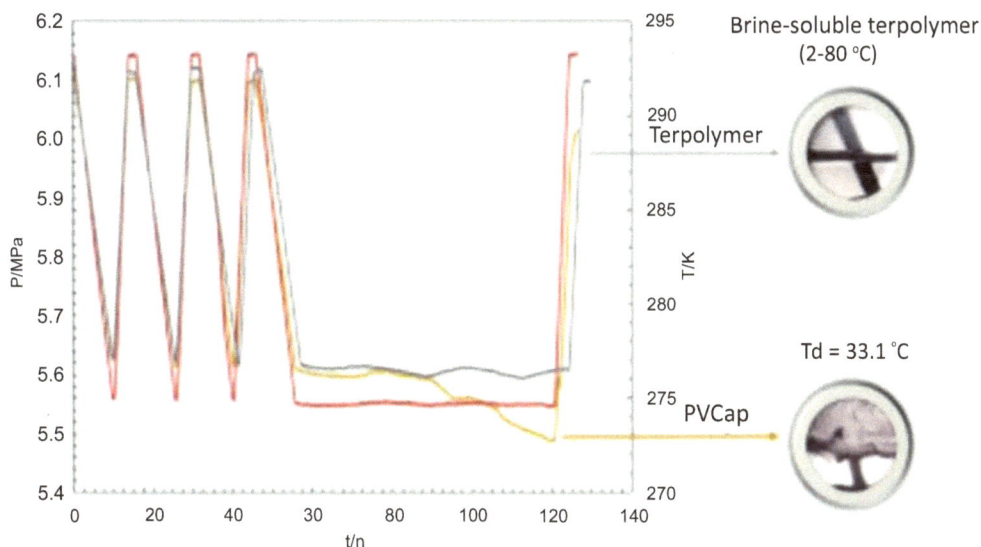

Fig. (7.5). Pressure and temperature variation as a function of time for type II natural gas hydrates in the presence of the terpolymer poly[VCapVBTMABrAA] and PVCap (polymer concentration: 0.5 g/100 g water).

ILs as Hydrate Formation Inhibitors

In recent years, ILs have also been studied as hydrate inhibitors. A good number of ILs with short alkyl chains have been described as LDHIs [28 - 30]. In this context, Del Villano and Kelland found that ILs with imidazolium as cation have limited performance compared to polymeric inhibitors [31], but have shown very good performance when forming synergistic mixtures with commercial LDIHs based on VCap polymers and hyperbranched ester polyamides [32 - 35].

Gupta *et al.* showed experimentally and theoretically that ILs containing aromatic (imidazolium-type) cations showed better performance as methane hydrate inhibitors than aliphatic (ammonium-type) cations [36]. However, it is known that certain quaternary ammonium and phosphonium salts, such as tetra(n-butyl)ammonium bromide (TBAB) and tetra(n-pentyl)ammonium bromide are able to disrupt efficiently the growth of type II structure hydrates [37 - 39].

Chua and Kelland showed that the IL tetra(iso-hexyl)ammonium bromide was an LDHI with excellent performance inhibiting tetrahydrofuran hydrates and its power was enhanced by forming a synergistic mixture with PVCap [40].

These ILs based on ammonium salts have also shown good performance as anti-agglomerant agents, for example, in 2013, Chua and Kelland studied the anti-caking effect of n-alkyl(n-butyl)ammonium bromides, finding that the effect can vary with chains between 8 and 18 carbon atoms, where the best prototype was the compound n-dodecyl-tri(n-butyl)-ammonium bromide; the anti-agglomerant effect also varied as a function of salinity. In this work, it was shown that there is an optimum alkyl chain length for maximum anti-agglomerant efficiency for this class of surfactant and that increasing salinity improves the anti-agglomerant performance. Performance is also improved by the addition of an anionic surfactant, such as sodium dodecyl sulfate (SDS), particularly at low salinity [41].

The excellent performance of different families of ILs as kinetic inhibitors of hydrate formation and, in several cases, also as anti-agglomerant agents, has been confirmed in several recent papers [33, 34, 42 - 45].

The following table describes some recent works, where ILs have been described as LDHIs.

Table 7.1. Some recent research works, where ILs have been described as LDHIs.

Studied ILs	Main Observations	Ref.
[EMIM]BF$_4$, [BMIM]BF$_4$, [EMIM]N(CN)$_2$, [EMIM]CF$_3$SO$_3$, [EMIM]EtSO$_4$	In this work, five ILs with imidazolium cation and different anions were studied. These ILs showed a dual thermokinetic function, *i.e.*, inhibitors that can not only change the behavior of the equilibrium hydrate dissociation curve (HLVE), but also slow down the nucleation and/or growth rate. Their effects on the HLVE curve were measured at pressures ranging from 30 to 110 bar while the effects on the induction time of hydrate formation were measured at -12°C and 114 bar. The 10 wt.% shifted the HLVE curve of methane hydrate by about 0.7-1.5 °C. The IL [BMIM]BF$_4$ showed the best performance because it exhibited the strongest hydrogen bond formation.	[46]
[EMIM]BF$_4$, [BMIM]BF$_4$	Because the experimental work in the previous paper was carried out at -12°C, giving very high subcooling (approx. -25°C), in this paper, we reinvestigated two of these ILs at more typical underwater temperatures and subcooling. We found that these ILs were very poor LDHIs when used alone at 5000-10000 ppm, but were quite good synergists for commercial LDHIs based on hyperbranched vinyl lactam and poly(esteramide) polymers. Both ILs showed only weak inhibition of tetrahydrofuran hydrate crystal growth.	[31]
N-(2-hydroxyethyl)-*N*-methylpyrrolidinium tetrafluoroborate, [HEMP]BF$_4$, 1-butyl-1-methylpyrrolidinium tetrafluoroborate [BMP]BF$_4$	It was shown that the two pyrrolidinium-based ILs designed for hydrate inhibition can greatly improve the induction time while changing the original equilibrium line. It was shown that the hydroxyl group on the chain of [HEMP]BF4 disrupted the formation of hydrogen bonds between water molecules. The possible application of dual-effect ILs, which not only change the equilibrium dissociation conditions, but also slow down the nucleation and crystal growth rates, was also shown.	[47]
[BMIM]BF$_4$, [BMIM]MeSO$_4$, [BMIM]HSO$_4$	The evaluated ILs showed to exert a dual effect on both methane hydrate formation and dissociation, including thermodynamic and kinetic inhibition. Kinetic modeling of methane hydrate inhibition by low-dose ILs was performed. The kinetic analysis showed that IL inhibitors mainly cause a delay in the nucleation or growth stage of the hydrate. The related inhibition mechanism was resolved with respect to the ionic nature and electrostatic interactions of ILs with water molecules. Two binomial exponential kinetic relationships were derived and used for simple methane hydrate formation in the presence of IL as LDHI.	[48]

Studied ILs	Main Observations	Ref.
[EMIM]Cl [EMIM]Br	The dissociation of methane hydrates was studied for solutions containing high concentrations of [EMIM]Cl up to 40 wt.% at 10, 15 and 20 MPa in a high-pressure micro differential scanning calorimeter, as well as the possible synergistic effect with sodium chloride (NaCl), monoethylene glycol (MEG), as well as an [EMIM]Cl/[EMIM]Br mixture was tested. The thermodynamic inhibition performance of the [EMIM]Cl and MEG mixtures showed a synergistic effect neither at low pressures nor at higher pressures. The [EMIM]Cl/[EMIM]Br mixture also exerted a synergistic effect at higher pressures. Unlike MEG or NaCl, inhibitors containing [EMIM]Cl or [EMIM]Br displayed an increase in inhibition efficiency as pressure increased.	[49]
EMIM]BF$_4$, N-butyl-N-methyl-Pyrrolidinium tetrafluoroborate [BMP]BF$_4$, N-(2-hydroxyethyl)-N-methylpyrrolidini-um tetrafluoroborate, [HEMP]BF$_4$	The concept of combining ILs with polymers used as KIHs to more effectively inhibit methane hydrate formation is described. The new inhibitors extended the induction time and decreased the hydrate growth rate. It was found that the presence of a hydroxyl group in the IL provided the most potent inhibition effect by forming hydrogen bonds between the IL and water molecules. The ILs were mixed with PVCap in a 1:1 wt.% ratio, preparing each mixture at 1% in water. In particular, the PVCap and [HEMP]BF4 mixture can significantly enhance the induction time by forming hydrogen bonds between the IL and water molecules.	[50]
1-ethyl-1-mehyl-pyrrolidinium tetrafluoroborate (TB) [EMP]BF$_4$, 1-butyl-1-methyl-pyrrolidinium TB [BMP]BF$_4$, 1-hexyl-1-methyl-pyrrolidinium TB [HMP]BF$_4$, 1-octyl-1-methyl-pyrrolidinium TB [OMP]BF$_4$, hydroxyethyl-1-methyl-pyrrolidinium TB, [HEMP]BF$_4$.	Experimental results revealed that the natural gas hydrate inhibition performance of ILs was much lower than that of PVCap when ILs were used alone. The PVCap-IL mixtures showed a dramatic increase in their hydrate inhibition efficiency, which was higher than that of PVCap alone. The PVCap-IL blend inhibited metastable sI-type hydrate nucleation. Additional hydrate inhibition experiments at higher pressure showed a reduced induction time of hydrate formation, which arose from a pressure-driven increase in the amount of gaseous reagent, inducing more hydrate nucleation sites. The mixtures [HMP]BF4 and PVCap-[HMP]BF4, exhibited the highest hydrate nucleation inhibition performance. The mixture of 1.0 wt.% PVCap and 0.5 wt.% HMP-BF4 exhibited the best LDHI performance within the tested composition range.	[32]

(Table 7.1) cont.....

Studied ILs	Main Observations	Ref.
1-hydroxyethyl-1-methylmorfolinium chloride [HEMM]Cl, 1-hydroxyethyl-1-methyl-morfolinium tetrafluoroborate [HEMM]BF$_4$.	The two ILs were synthesized and the different effects of the anions on the kinetics of methane hydrate formation were investigated. HEMM]Cl and [HEMM]BF4 acted as kinetic hydrate promoter and inhibitor, respectively. The induction time of [HEMM]BF4 solutions increased in proportion to the concentration of [HEMM]BF4, and both ILs showed a thermodynamic effect on methane hydrate inhibition. The X-ray diffraction pattern of hydrates formed in the presence of both ILs showed that there was no influence or incorporation of ILs into the [HEMM]Cl crystal structure, which could distort the rigid hydrate host structure on the surface by hydrogen bonds between ions and water molecules, promoting the penetration or inclusion of methane into the growing clathrate hydrate structures, thus enhancing hydrate formation. Conversely, the kinetic inhibition behavior of [HEMM]BF4 could be attributed to the hypothesis that BF4 could act as a mobile pseudo-guest.	[51]
Tetraalkylammonium acetate TMAA, Choline butyrate [Ch-But], Choline Isobutyrate [Ch-iB], Choline Hexanoate [Ch-Hex], Choline Octanoate (Ch-Oct).	A systematic study of methane hydrate inhibition in the presence of a five-ammonium-cation family was carried out. Hydrate equilibrium curves were obtained for methane in pure water and in 1 and 5 wt.% aqueous solutions of the ILs. The efficiency of thermodynamic hydrate inhibition and promotion in each IL was thoroughly investigated by observing the changes in the hydrate equilibrium curves and the trends of calculated hydrate suppression temperatures. Results were also obtained for methanol to validate the apparatus and compare the inhibition efficiency of the ILs. Molar hydrate dissociation enthalpies were obtained for the studied systems using the Clausius-Clapeyron equation and showed that the ILs did not participate in the formation of hydrate cages. The induction time data revealed that some of the studied ILs delayed the hydrate formation time, indicating a possible kinetic inhibitory effect. The tested ILs exhibited a variety of phenomena, such as hydrate inhibition, hydrate stabilization, kinetic effect and surfactant-like behavior.	[52]

(Table 7.1) cont.....

Studied ILs	Main Observations	Ref.
1-(2-hydroxyethyl)-3-methylimidazolium chloride [OH-EMIM]Cl, 1-(2-hydroxyethyl)-3-methylimidazolium bromide [OH-EMIM]Br 1-butyl-3-methylimidazolium bromide [BMIM]Br, 1-butyl-3-methylimidazolium chloride [BMIM]Cl, 1-butyl-3-methylimidazolium perchlorate [BMIM]ClO$_4$, 1-butyl-3-methylimidazolium dicyanamide [BMIM]N(CN)$_2$, 1-butyl-3-methylimidazolium hydrogensulfate, [BMIM]HSO$_4$ 1-butyl-3-methylimidazolium trifluoromethanesulfonate, [BMIM]CF$_3$SO$_3$ 1-butyl-3-methylimidazolium methylsulfate, [BMIM]CH$_3$SO$_4$	The performance of nine ILs as LDHIs for methane gas hydrates was investigated employing a high-pressure micro differential scanning calorimeter. Aqueous solutions of IL (0.01 w/w) as well as polyvinylpyrrolidone (PVP) were prepared, and the induction time of methane hydrate formation in these solutions was measured at 7.1 MPa and 258.15 K. It was found that [BMIM]CF$_3$SO$_3$, [BMIM]CH$_3$SO$_4$ and [OH-EMIM]Br could retard the hydrate formation at this concentration. Their relative inhibition power (RIP) was higher than that of PVP. The other ILs showed shorter induction times when compared with the blank sample; a promotional effect rather than an inhibition effect on hydrate formation was identified. It was found that there was a strong correlation between the molar mass of [BMIM]-based ILs and the induction time. The Avrami analysis indicated that most of the methane hydrate crystallization process was governed by the clathrate formation reaction and slightly driven by the diffusion-controlled mechanism.	[53]

Organic surfactants have shown to play a very important role preventing gas hydrate formation by delaying hydrate nucleation or crystal growth or by preventing hydrate agglomeration and deposition [54]. Moreover, recent works have evidenced the potential of long-chain ILs as LDHIs. In this study, 1-dodecy--3-methylimidazolium tetrafluoroborate was used and compared with a widely used large-scale LDHI, such as PVCap, and it was found that these ILs with surfactant properties were very effective performing hydrate inhibition under the same experimental conditions as the commercial product and in addition, these ILs also showed anti-agglomerating properties [55].

Recently, the IMP developed a pyridinium cation-based IL with a 12-carbon chain and bromide anion, which also exhibited surfactant properties. This compound showed excellent results as LDHI with PVCap-like performance under the studied experimental conditions. In the article, Synthesis and evaluation of a pyridinium-based-ionic-liquid-type surfactant as a new low-dosage-methane-hydrate inhibitor (Energy & Fuels 2020, 34, 1706-1715) by Ascención Romero-Martínez, Erika Hernández-Guerrero, Edgar Ramírez-Jaramillo and Rafael Martínez-Palou [56], an IL with pyridinium cation as LDHI was studied for the first time and it was shown that this compound also presented excellent anti-agglomerant performance. This IL was synthesized quickly and efficiently with the use of microwaves. (MW, Fig. **7.5**).

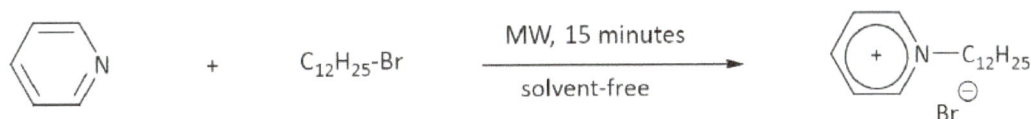

Fig. (7.5). Microwave-assisted synthesis of C12PyBr.

This pyridinium cation had not been previously described as LDHI and this pyridinium family is cheaper than the most studied one corresponding to imidazolium; on the other hand, as it can be seen in Fig. (7.5), it is possible to obtain the product in only 15 min of reaction without using solvents. Additionally, this compound presents an alkyl chain of 12 carbon atoms that allows it to work as a surfactant, so that the compound shows high performance as an LDHI and as anti-agglomerant agent.

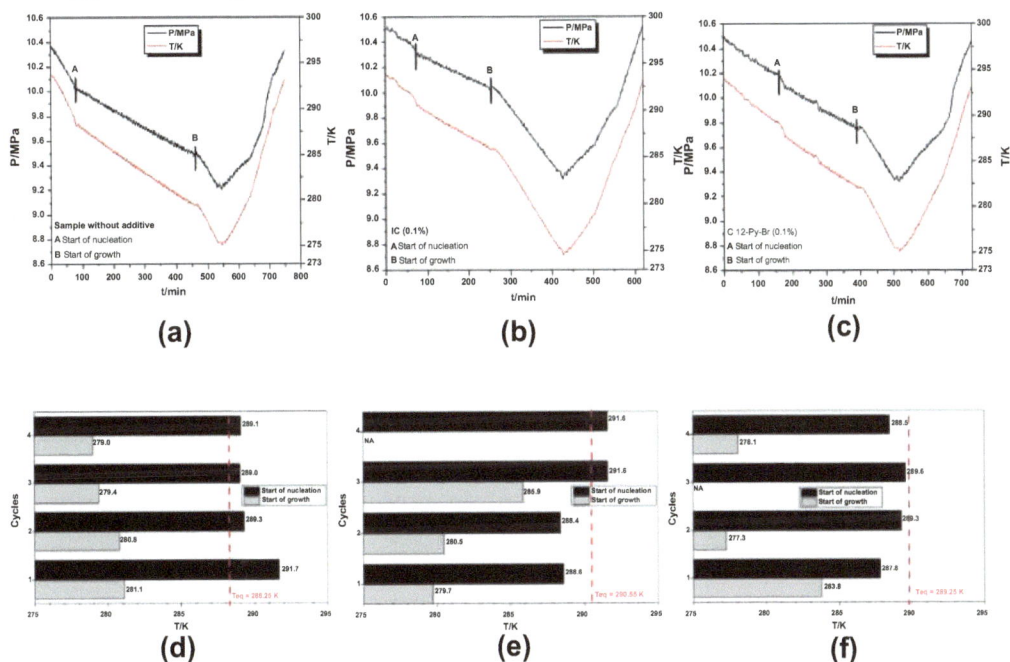

Fig. (7.6). Pressure and Temperature *vs.* time curves obtained for: **(a)** sample without additive, **(b)** with Inhibex 101TM (0.1 wt.%) and **(c)** using C12PyBr (0.1 wt.%); and: **(A)** temperature at the beginning of hydrate nucleation, **(B)** temperature at the beginning of the crystal growth stage, for: **(d)** sample without additive, **(e)** with Inhibex 101TM (0.1 wt.%) and **(f)** with C12PyBr (0.1 wt.%).

Fig. (7.6) shows the pressure and temperature *versus* time curves obtained for: (a) the sample without additive, (b) with Inhibex 101TM (this inhibitor is PVCap based) at 0.1% and (c) C12PyBr at 0.1%. In addition, in the same figure, the results of the evaluation of the efficiency of the inhibitors at different concentrations (0.1, 0.5 and 1.0%) are presented. In this case, the temperature values at the beginning of nucleation and hydrate growth are displayed for the 0.1% concentration for: (d) sample without additive, (e) using Inhibex 101TM and (f) using C12PyBr. For each evaluation, 4 cycles of experiments were performed.

According to the analysis of the results, C12PyBr at 1.0% shows performance comparable to that of the commercial inhibitor evaluated in this study, and its performance was higher when applied at a lower concentration (0.1%) compared to the situation when its concentration was 1.0% in mass, with respect to water; this fact could be explained by its surfactant properties and the ability to form hydrogen bonds, and also, perhaps this behavior was due to the fact that 0.1% was the optimal inhibitory concentration, and from this point, higher concentrations generate an overdose that has a negative effect on the inhibitory process.

Considering the cost, the simplicity of its synthesis procedure and its efficiency as LDHI, C12PyBr could be a suitable candidate for a future industrial application as LDHI.

Dual-purpose ILs with Simultaneous LDHIs and Corrosion Inhibitors

In the last decade, the concept of LDHIs also considers the simultaneous work as corrosion inhibitors (CIs), which has attracted great attention for deepwater flow assurance since both problems (hydrate formation and deposition and corrosion) are very common in hydrocarbon extraction in deepwater wells and with this type of inhibitors, chemical expenses and infrastructure demands can be considerably reduced [26]. In a recent review on this topic, some imidazolium-type ILs were described as very effective hydrate and Cis [57].

Several long-chain alkyl pyridinium bromide compounds have also been described as effective CIs of steel in acidic media [58 - 60], so it is to be expected that the C12PyBr hydrate inhibitor described by the IMP and discussed earlier in this chapter may act as a dual inhibitor with great prospect for industrial application; however, although ILs are considered environmentally friendly compounds mainly because of their zero vapor pressure, it is important to carry out a detailed toxicity study to evaluate how environmentally friendly these compounds really are [61].

CONCLUDING REMARKS

In this chapter, some theoretical aspects related to the formation of gas hydrates characteristic of oil and gas extraction in deep waters were discussed. As the literature review on the subject shows, ILs play an important role not only as chemicals with hydrate inhibitory properties but also simultaneously acting as corrosion inhibitors. On the other hand, recent studies show that ILs, most of all imidazolium and benzyltrimethylammonium derivatives, can form part of the terpolymer structure with VCap, and show excellent performance at 6 MPa pressure and 1 °C for this application, functioning as LDHIs. The mixture of ILs with polymers delays the hydrate formation time and enhances the effect of poly(vinyl caprolactam).

REFERENCES

[1]   Sloan, E.D.; Koh, C.A. *Clathrate Hydrates of Natural Gas*; Taylor & Francis Group: New York, **2008**.

[2]   Kelland, M.A. Gas Hydrate Control. In: *Production chemicals for the oil and gas industry*; CRC Press, **2014**.
       [http://dx.doi.org/10.1201/b16648-13]

[3]   Kelland, M.A. History of the Development of Low Dosage Hydrate Inhibitors. *Energy Fuels,* **2006**, *20*(3), 825-847.
       [http://dx.doi.org/10.1021/ef050427x]

[4]   Carroll, J. Natural Gas Hydrates: A Guide for Engineers. Gulf Professional Publishing: Waltham, **2014**.

[5]   Tohidi, B.; Anderson, R.; Mozaffar, H.; Tohidi, F. The return of kinetic hydrate inhibitors. *Energy Fuels,* **2015**, *29*(12), 8254-8260.
       [http://dx.doi.org/10.1021/acs.energyfuels.5b01794]

[6]   Perrin, A.; Musa, O.M.; Steed, J.W. The chemistry of low dosage clathrate hydrate inhibitors. *Chem. Soc. Rev.,* **2013**, *42*(5), 1996-2015.
       [http://dx.doi.org/10.1039/c2cs35340g] [PMID: 23303391]

[7]   Kelland, M.A.A. *Review of kinetic hydrate inhibitors - Tailormade water-soluble polymers for oil and gas industry applications. Advances in Materials Science Research*; Nova Science Publishers: New York, **2011**.

[8]   Kelland, M.A. A Review of Kinetic Hydrate Inhibitors from an Environmental Perspective. *Energy Fuels,* **2018**, *32*(12), 12001-12012.
       [http://dx.doi.org/10.1021/acs.energyfuels.8b03363]

[9]   Kamal, M.S.; Hussein, I.A.; Sultan, A.S.; von Solms, N. Application of various water soluble polymers in gas hydrate inhibition. *Renew. Sustain. Energy Rev.,* **2016**, *60*, 206-225.
       [http://dx.doi.org/10.1016/j.rser.2016.01.092]

[10]   Masri, A.N.; Sulaimon, A.A. Amino acid-based ionic liquids as dual kinetic-thermodynamic methane hydrate inhibitor. *J. Mol. Liq.,* **2022**, *349*, 118481.
       [http://dx.doi.org/10.1016/j.molliq.2022.118481]

[11]   Rivers, G.T.; Crosby, D.L. Method and compositions for inhibiting formation of hydrocarbon hydrates. *International Patent Application WO2004/022909,* **2004**.

[12]   Leinweber, D.; Feustel, M.U.S. Use of Polyesters in the Form of Gas Hydrate Inhibitors. *Patent Application 20080214865.*

[13]   Leinweber, D.; Roesch, A.R.; Feustel, M. Development of A New Class of Green Kinetic Hydrate Inhibitors. *U.S. Pat. Appl. 20090054268,* **2009**.

[14]   Cole, R.A.; Grinrod, A.; Cely, A. *Kinetic hydrate inhibitors with pendent amino functionality.,* Int. Pat. Appl. *WO/2014/078163,* **2004**.

[15]   Gonzáles, R.; Djuve, J. *Hydrate inhibitors.,* Int. Pat. Appl. *WO/2010/101477,* **2010**.

[16]   Musa, O.M.; Cuiyue, L. *Degradable polymer compositions and uses thereof.,* Int. Pat. Appl. *WO/2010/114761,* **2010**.

[17]   Klomp, U. C. Method for inhibiting the plugging of conduits by gas hydrates. *Int. Pat. Appl. WO/2013/096201,* **2013**.

[18]   Clements, J.; Pakulski, M. K.; Riethmeyer, J.; Lewis, D. C. Improved poly(vinyl caprolactam) kinetic gas hydrate inhibitor and method for preparing the same. *Canada Pat. CA2980007A1.*

[19]   Carlise, J. R.; Lindeman, O. E. S.; Reed, P. E.; Conrad, P. G.; Ver Vers, L. M. Method of controlling gas hydrates in fluid systems. *Austrlian Pat. AU2018250400B2.*

[20]   Yuan, J.; Mecerreyes, D.; Antonietti, M. Poly(ionic liquid)s: An update. *Prog. Polym. Sci.,* **2013,** *38*(7), 1009-1036.
[http://dx.doi.org/10.1016/j.progpolymsci.2013.04.002]

[21]   MacFarlane, D.R.; Forsyth, M.; Howlett, P.C.; Kar, M.; Passerini, S.; Pringle, J.M.; Ohno, H.; Watanabe, M.; Yan, F.; Zheng, W.; Zhang, S.; Zhang, J. Ionic liquids and their solid-state analogues as materials for energy generation and storage. *Nat. Rev. Mater.,* **2016,** *1*(2), 15005.
[http://dx.doi.org/10.1038/natrevmats.2015.5]

[22]   Wang, X.; Zhu, H.; Girard, G.M.A.; Yunis, R.; MacFarlane, D.R.; Mecerreyes, D.; Bhattacharyya, A.J.; Howlett, P.C.; Forsyth, M. Preparation and characterization of gel polymer electrolytes using poly(ionic liquids) and high lithium salt concentration ionic liquids. *J. Mater. Chem. A Mater. Energy Sustain.,* **2017,** *5*(45), 23844-23852.
[http://dx.doi.org/10.1039/C7TA08233A]

[23]   Qian, W.; Texter, J.; Yan, F. Frontiers in poly(ionic liquid)s: syntheses and applications. *Chem. Soc. Rev.,* **2017,** *46*(4), 1124-1159.
[http://dx.doi.org/10.1039/C6CS00620E] [PMID: 28180218]

[24]   Eftekhari, A.; Saito, T. Synthesis and properties of polymerized ionic liquids. *Eur. Polym. J.,* **2017,** *90,* 245-272.
[http://dx.doi.org/10.1016/j.eurpolymj.2017.03.033]

[25]   Nakarit, C.; Kelland, M.A.; Liu, D.; Chen, E.Y.X. Cationic kinetic hydrate inhibitors and the effect on performance of incorporating cationic monomers into N-vinyl lactam copolymers. *Chem. Eng. Sci.,* **2013,** *102,* 424-431.
[http://dx.doi.org/10.1016/j.ces.2013.06.054]

[26]   Rebolledo-Libreros, M.E.; Reza, J.; Trejo, A.; Guzmán-Lucero, D.J. Evaluation of copolymers from 1-vinyl-3-alkylimidazolium bromide and N -vinylcaprolactam as inhibitors of clathrate hydrate formation. *J. Nat. Gas Sci. Eng.,* **2017,** *40,* 114-125.
[http://dx.doi.org/10.1016/j.jngse.2017.02.008]

[27]   Reza, J.; Trejo, A.; Rebolledo-Libreros, M.E.; Guzmán-Lucero, D. Inhibition of structure II hydrates formation by salt-tolerant N-vinyl lactam-based terpolymers. *J. Nat. Gas Sci. Eng.,* **2018,** *56,* 175-192.
[http://dx.doi.org/10.1016/j.jngse.2018.05.039]

[28]   Tariq, M.; Rooney, D.; Othman, E.; Aparicio, S.; Atilhan, M.; Khraisheh, M. Gas Hydrate Inhibition: A Review of the Role of Ionic Liquids. *Ind. Eng. Chem. Res.,* **2014,** *53*(46), 17855-17868.
[http://dx.doi.org/10.1021/ie503559k]

[29]   Menezes, D.É.S.; Pessôa Filho, P.A.; Robustillo Fuentes, M.D. Use of 1-butyl-3-methylimidazol-um-based ionic liquids as methane hydrate inhibitors at high pressure conditions. *Chem. Eng. Sci.,* **2020,** *212,* 115323.
[http://dx.doi.org/10.1016/j.ces.2019.115323]

[30]   Ul Haq, I.; Qasim, A.; Lal, B.; Zaini, D.B.; Foo, K.S.; Mubashir, M.; Khoo, K.S.; Vo, D.V.N.; Leroy, E.; Show, P.L. Ionic liquids for the inhibition of gas hydrates. A review. *Environ. Chem. Lett.,* **2022,** *20*(3), 2165-2188.
[http://dx.doi.org/10.1007/s10311-021-01359-9]

[31]   Del Villano, L.; Kelland, M.A. An investigation into the kinetic hydrate inhibitor properties of two imidazolium-based ionic liquids on Structure II gas hydrate. *Chem. Eng. Sci.,* **2010,** *65*(19), 5366-

5372.
[http://dx.doi.org/10.1016/j.ces.2010.06.033]

[32]   Lee, W.; Shin, J.Y.; Kim, K.S.; Kang, S.P. Synergetic Effect of Ionic Liquids on the Kinetic Inhibition Performance of Poly(*N* -vinylcaprolactam) for Natural Gas Hydrate Formation. *Energy Fuels,* **2016**, *30*(11), 9162-9169.
[http://dx.doi.org/10.1021/acs.energyfuels.6b01830]

[33]   Lee, W.; Shin, J.Y.; Cha, J.H.; Kim, K.S.; Kang, S.P. Inhibition effect of ionic liquids and their mixtures with poly(M *N*-Vinylcaprolactam) on methane hydrate formation. *J. Ind. Eng. Chem.,* **2016**, *38*, 211-216.
[http://dx.doi.org/10.1016/j.jiec.2016.05.007]

[34]   Lee, W.; Kim, K.S.; Kang, S.P.; Kim, J.N. Synergetic Performance of the Mixture of Poly(*N* -vinylcaprolactam) and a Pyrrolidinium-Based Ionic Liquid for Kinetic Hydrate Inhibition in the Presence of the Mineral Oil Phase. *Energy Fuels,* **2018**, *32*(4), 4932-4941.
[http://dx.doi.org/10.1021/acs.energyfuels.8b00294]

[35]   Kang, S-P.; Jung, T.; Lee, J.W. Macroscopic and spectroscopic identifications of the synergetic inhibition of an ionic liquid on hydrate formations. *Chem. Eng. Sci.,* **2016**, *143*, 270-275.
[http://dx.doi.org/10.1016/j.ces.2016.01.009]

[36]   Gupta, P.; Sakthivel, S.; Sangwai, J.S. Effect of aromatic/aliphatic based ionic liquids on the phase behavior of methane hydrates: Experiments and modeling. *J. Chem. Thermodyn.,* **2018**, *117*, 9-20.
[http://dx.doi.org/10.1016/j.jct.2017.08.037]

[37]   Koh, C.A.; Westacott, R.E.; Zhang, W.; Hirachand, K.; Creek, J.L.; Soper, A.K. Mechanisms of gas hydrate formation and inhibition. *Fluid Phase Equilib.,* **2002**, *194-197*, 143-151.
[http://dx.doi.org/10.1016/S0378-3812(01)00660-4]

[38]   Mady, M.F.; Kelland, M.A. Synergism of *tert* -Heptylated Quaternary Ammonium Salts with Poly(*N* -vinyl caprolactam) Kinetic Hydrate Inhibitor in High-Pressure and Oil-Based Systems. *Energy Fuels,* **2018**, *32*(4), 4841-4849.
[http://dx.doi.org/10.1021/acs.energyfuels.8b00110]

[39]   Norland, A.K.; Kelland, M.A. Crystal growth inhibition of tetrahydrofuran hydrate with bis- and polyquaternary ammonium salts. *Chem. Eng. Sci.,* **2012**, *69*(1), 483-491.
[http://dx.doi.org/10.1016/j.ces.2011.11.003]

[40]   Chua, P.C.; Kelland, M.A. Tetra(iso-hexyl)ammonium Bromide—The Most Powerful Quaternary Ammonium-Based Tetrahydrofuran Crystal Growth Inhibitor and Synergist with Polyvinylcaprolactam Kinetic Gas Hydrate Inhibitor. *Energy Fuels,* **2012**, *26*(2), 1160-1168.
[http://dx.doi.org/10.1021/ef201849t]

[41]   Chua, P.C.; Kelland, M.A. Study of the Gas Hydrate Anti-agglomerant Performance of a Series of *n* -Alkyl-tri(*n* -butyl)ammonium Bromides. *Energy Fuels,* **2013**, *27*(3), 1285-1292.
[http://dx.doi.org/10.1021/ef3018546]

[42]   Ke, W.; Chen, D. A short review on natural gas hydrate, kinetic hydrate inhibitors and inhibitor synergists. *Chin. J. Chem. Eng.,* **2019**, *27*(9), 2049-2061.
[http://dx.doi.org/10.1016/j.cjche.2018.10.010]

[43]   Liu, Y.; Wang, X.; Lang, C.; Zhao, J.; Lv, X.; Ge, Y.; Jiang, L. Experimental study on the gas hydrates blockage and evaluation of kinetic inhibitors using a fully visual rocking cell. *J. Nat. Gas Sci. Eng.,* **2021**, *96*, 104331.
[http://dx.doi.org/10.1016/j.jngse.2021.104331]

[44]   Hussain, H.H.; Husin, H. Review on Application of Quaternary Ammonium Salts for Gas Hydrate Inhibition. *Appl. Sci. (Basel),* **2020**, *10*(3), 1011.
[http://dx.doi.org/10.3390/app10031011]

[45]   Wang, Y.; Fan, S.; Lang, X. Reviews of gas hydrate inhibitors in gas-dominant pipelines and

application of kinetic hydrate inhibitors in China. *Chin. J. Chem. Eng.,* **2019**, *27*(9), 2118-2132.
[http://dx.doi.org/10.1016/j.cjche.2019.02.023]

[46]   Xiao, C.; Adidharma, H. Dual function inhibitors for methane hydrate. *Chem. Eng. Sci.,* **2009**, *64*(7), 1522-1527.
[http://dx.doi.org/10.1016/j.ces.2008.12.031]

[47]   Kim, K.S.; Kang, J.W.; Kang, S.P. Tuning ionic liquids for hydrate inhibition. *Chem. Commun. (Camb.),* **2011**, *47*(22), 6341-6343.
[http://dx.doi.org/10.1039/c0cc05676f] [PMID: 21547283]

[48]   Nazari, K.; Moradi, M.R.; Ahmadi, A.N. Kinetic Modeling of Methane Hydrate Formation in the Presence of Low-Dosage Water-Soluble Ionic Liquids. *Chem. Eng. Technol.,* **2013**, *36*(11), 1915-1923.
[http://dx.doi.org/10.1002/ceat.201300285]

[49]   Richard, A.R.; Adidharma, H. The performance of ionic liquids and their mixtures in inhibiting methane hydrate formation. *Chem. Eng. Sci.,* **2013**, *87*, 270-276.
[http://dx.doi.org/10.1016/j.ces.2012.10.021]

[50]   Kang, S.P.; Kim, E.S.; Shin, J.Y.; Kim, H.T.; Kang, J.W.; Cha, J.H.; Kim, K.S. Unusual synergy effect on methane hydrate inhibition when ionic liquid meets polymer. *RSC Advances,* **2013**, *3*(43), 19920-19923.
[http://dx.doi.org/10.1039/c3ra43891k]

[51]   Lee, W.; Shin, J.Y.; Kim, K.S.; Kang, S.P. Kinetic Promotion and Inhibition of Methane Hydrate Formation by Morpholinium Ionic Liquids with Chloride and Tetrafluoroborate Anions. *Energy Fuels,* **2016**, *30*(5), 3879-3885.
[http://dx.doi.org/10.1021/acs.energyfuels.6b00271]

[52]   Tariq, M.; Connor, E.; Thompson, J.; Khraisheh, M.; Atilhan, M.; Rooney, D. Doubly dual nature of ammonium-based ionic liquids for methane hydrates probed by rocking-rig assembly. *RSC Advances,* **2016**, *6*(28), 23827-23836.
[http://dx.doi.org/10.1039/C6RA00170J]

[53]   Nashed, O.; Sabil, K.M.; Ismail, L.; Japper-Jaafar, A.; Lal, B. Mean induction time and isothermal kinetic analysis of methane hydrate formation in water and imidazolium based ionic liquid solutions. *J. Chem. Thermodyn.,* **2018**, *117*, 147-154.
[http://dx.doi.org/10.1016/j.jct.2017.09.015]

[54]   Phillips, N.J.; Kelland, M.A. The Application of Surfactants in Preventing Gas Hydrate Formation. In: *Industrial Applications of Surfactants IV*; Woodhead Publishing, **1999**; pp. 244-259.
[http://dx.doi.org/10.1533/9781845698614.244]

[55]   Saikia, T.; Mahto, V. Evaluation of 1-Decyl-3-Methylimidazolium Tetrafluoroborate as clathrate hydrate crystal inhibitor in drilling fluid. *J. Nat. Gas Sci. Eng.,* **2016**, *36*, 906-915.
[http://dx.doi.org/10.1016/j.jngse.2016.11.029]

[56]   Romero-Martínez, A.; Hernández-Guerrero, E.; Ramírez-Jaramillo, E.; Martínez-Palou, R. Synthesis and evaluation of a pyridinium-based-ionic-liquid-type surfactant as new low-dosage-methane-hydrate inhibitor. *Energy Fuels,* **2020**, *34*(8), 9243-9251.
[http://dx.doi.org/10.1021/acs.energyfuels.0c00426]

[57]   Qasim, A.; Khan, M.S.; Lal, B.; Shariff, A.M. A perspective on dual purpose gas hydrate and corrosion inhibitors for flow assurance. *J. Petrol. Sci. Eng.,* **2019**, *183*, 106418.
[http://dx.doi.org/10.1016/j.petrol.2019.106418]

[58]   Olivares-Xometl, O.; López-Aguilar, C.; Herrastí-González, P.; Likhanova, N.V.; Lijanova, I.; Martínez-Palou, R.; Rivera-Márquez, J.A. Adsorption and corrosion inhibition performance by three new ionic liquids on API 5L X52 steel surface in acid media. *Ind. Eng. Chem. Res.,* **2014**, *53*(23), 9534-9543.
[http://dx.doi.org/10.1021/ie4035847]

[59]   Xia, G.; Jiang, X.; Zhou, L.; Liao, Y.; duan, M.; Wang, H.; Pu, Q.; Zhou, J. Synergic effect of methyl acrylate and N-cetylpyridinium bromide in N-cetyl-3-(2-methoxycarbonylvinyl)pyridinium bromide molecule for X70 steel protection. *Corros. Sci.,* **2015**, *94*, 224-236.
       [http://dx.doi.org/10.1016/j.corsci.2015.02.005]

[60]   Ben Aoun, S. On the corrosion inhibition of carbon steel in 1 M HCl with a pyridinium-ionic liquid: chemical, thermodynamic, kinetic and electrochemical studies. *RSC Advances,* **2017**, *7*(58), 36688-36696.
       [http://dx.doi.org/10.1039/C7RA04084A]

[61]   Haq, I.U.; Qasim, A.; Lal, B.; Zaini, D.B. Mini review on environmental issues concerning conventional gas hydrate inhibitors. *Process Saf. Prog.,* **2022**, *41*(S1), S129-S134.
       [http://dx.doi.org/10.1002/prs.12325]

CHAPTER 8

ILs Applied to Enhance Oil Recovery Processes

Abstract: The application of chemical products, particularly polymeric products, is undoubtedly one of the most helpful and effective alternatives for EOR processes and changes in oil mobility. Appropriate chemical products provide more favorable interfacial conditions to the flow of petroleum, reducing the interfacial tensions between water and oil, and therefore, increasing the miscibility of these two compounds, and allowing the oil to flow in the porous medium. In this context, ionic compounds and particularly ILs are proving to be important auxiliaries in the performance of chemicals used for this application, particularly in extreme conditions of temperature and salinity.

Keywords: Ammonium, Anionic, Cationic, Tertiary recovery, Enhanced Oil Recovery, High salinity, Interfacial tension, Ionic liquids, Miscibility, Permeability, Switterionic, Vinylpyrrolidone, Wettability alteration.

INTRODUCTION

In the hydrocarbon market, at the international level, Mexico has diminished its participation due to the decrease in the production of fossil fuels at the national level in such a way that in 2013, the production reached 2.52 mbpd; in 2015, 2.26 mbpd; in 2017, 1.94 mbpd; in 2018, 1.83 mbpd; and in July, 1.69 mbpd. From 2014, because of the fall in the price of hydrocarbons, the inversion in the exploration at global level diminished 60% and PEMEX reduced its investment to 37%. Although the exploration and production activities are regulated, it is important that projects ensuring the sustainability of the Mexican energy industry be developed. In this sense, the production projects in conventional reservoirs of mature fields are a good option to keep the Mexican oil industry in the black. From the last century (1950s) up to date, Mexico had several secondary recovery [1 - 3] projects, mainly those featuring the injection of water into the fields: San Andrés in 1961; Tamaulipas-Constitutions in 1968; El Golpe, La Venta, Antonio J. Bermúdez, Sitio Grande, Cuichapa, Rodador and Magallanes in the years 1970-1978; Ogarrio in 1983; Abkatún-Pol-Chuc in 1991, *etc*. Also, the injection of nitrogen into the reservoir Cantarell and thermal recovery through vapor at the Samaria field has been carried out as part of several projects. As for the tertiary oil recovery, Mexico practically avoided the injection of chemical compounds (surfactants and polymers), while in other countries, this practice is very common and only restricted by the relationship between the oil price and cost of the injected chemical.

About 2.0×1012 barrels (0.3×1012 m^3) of conventional oil and 5.0×1012 barrels (0.8×1012 m^3) of heavy oil will remain in reservoirs around the world once the conventional recovery methods become exhausted. It should be kept in mind that the tertiary oil recovery methods can reach recovery factors above 60% whereas surfactant and polymer flooding belongs to well-known enhanced recovery methods that imply low risk and application under wide reservoir conditions. The aim of injecting polymers is to improve the sweeping efficiency in the reservoir and diminish the mobility contrast between water and oil when oil has a higher viscosity than that of the injected water. To date, there are more than 150 references worldwide and 30 years of experience in polymer flooding, however, the method employing surfactants is even more popular among the residual oil recovery processes, for it can be used at low surfactant concentrations, thus provoking important diminution of the interfacial tension (IFT) between oil and water [4]. Frequently, different chemical injection ways, including those featuring binary mixtures such as alkali/surfactant (AS), surfactant/polymer (SP), alkali/polymer (AP) or alkali/surfactant/polymer (ASP) have been used in tertiary recovery processes. The residual oil saturation and recovery process can be described indirectly by means of the capillary number (Nc), which is defined by the equation (1).

$$Nc = \frac{\mu v}{\sigma \cos \theta},$$ (1)

where μ is the viscosity of the displacing fluid, v is the Darcy displacement velocity, $\cos \theta$ is the contact angle and σ is the IFT between the displacing (water) and displaced (oil) fluids. Higher Nc means higher oil recovery; then, after the secondary oil recovery process by water injection, Nc ranges from $10-7$ to $10-6$ whereas with tertiary recovery processes, Nc can reach up to $10-2$, thus reducing the oil residual saturation to the minimum. The increase in the viscosity of the displacing fluid (μ) using polymer solutions augments the number of capillaries less than 100 times while the application of surfactants to reduce the IFT (σ) can increase Nc 1000 times [5 - 6].

Currently, there are four types of surfactants: anionic, non-ionic, cationic and zwitterionic (Fig. **8.1**). ILs represent a particular case of the cationic or zwitterionic surfactants, for they are liquid salts (at ambient temperature) with organic cation.

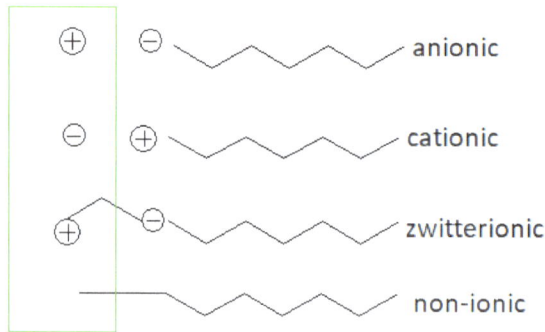

Fig. (8.1). General structure of surfactants.

Cationic surfactants are capable of desorbing oil irreversibly from the rock surface through the formation of ion pairs between the surfactant and carboxylic groups of crude oil naphthenic acids; for this reason, cationic surfactants are better agents of the wettability change than the non-ionic or anionic surfactants. There is a comparative study featuring an anionic surfactant derived from alkyl benzene sulfonic acid and the cationic surfactant cetyltrimethylammonium bromide (CTAB), where the authors showed that although the effect exerted by the cationic surfactant diminishing the IFT was more efficient than that displayed by the anionic one, it could recover up to 55% of crude oil while CTAB recovered as much as 70% [7].

The potential of the mixed system of cationic surfactants 1-dodecyl-3-methylimidazolium chloride and sodium dodecyl sulfate was evaluated through the analysis of IFT reduction, emulsion stability, wettability alteration and additional oil recovery tests by rock core injection, identifying a synergistic effect, which can be attributed to the formation of pseudo-bidirectional systems by electrostatic attraction (Fig. **8.2**). In addition, the mixed systems exhibited emulsion stability and excellent tolerance to salt and temperature with the IFT reduction, accomplishing additional oil recovery up to 17% in the presence of 50 000 mg / L of NaCl and temperature equal to 85 °C [8].

Most papers devoted to ILs for EOR processes have been focused on studies considering the IFT reduction or rock wettability change, however, until now, all the works related to EOR processes have been carried out at laboratory level [9]. In general, in comparison with non-ionic and anionic surfactants, ILs are more efficient changing the IFT under high salinity conditions [10, 11]. By increasing the length of the side chain of the cationic part, the effect on the diminution of the IFT becomes more evident, producing the lowest values for 1-alkyl-

3-methylimidazolium chloride with alkyl chain length between C_{12} - C_{18}; in addition, imidazolium-based ILs are more efficient than those based on pyridinium [12 - 15]. The same behavior pattern was discovered for the 1-akyl-3-methylimidazolium alkyl sulfates $[C_n mim][CmSO_4]$ [16] and $[C_n mim][BF_4]$ [17], where the length increase of the alkyl chain in either the cation or anion can lead to both lower critical micelle concentration (CMC) values and higher efficiency reducing the surface tension; even the addition of an extra methylene group to the longest alkyl chain exerts a more important effect on the surface activity than by adding it to the shortest one, reaching values of 0.03 mN/m for compounds with chain lengths consisting of 12 carbon atoms. With imidazolium-based-gemini ILs, $[C_{12}im-C_4-imC_{12}]Br_2$, it was possible to reduce the IFT drastically up to $8,2 \times 10^{-3}$ mN (reduction of 99,97%) using IL concentrations below 0,2 mol/L, in addition to form a stable emulsion with crude oil [18]. The combination of the IL derived from 1-alkyl-3methylimidazolium bromide and hydrophobically modified polymer was capable of reducing the IFT even under high salinity and temperature conditions, showing excellent viscosity behavior (22.2 at 17.4 mPa) from 30 to 90 °C, accomplishing additional oil recovery of 21.6% [19].

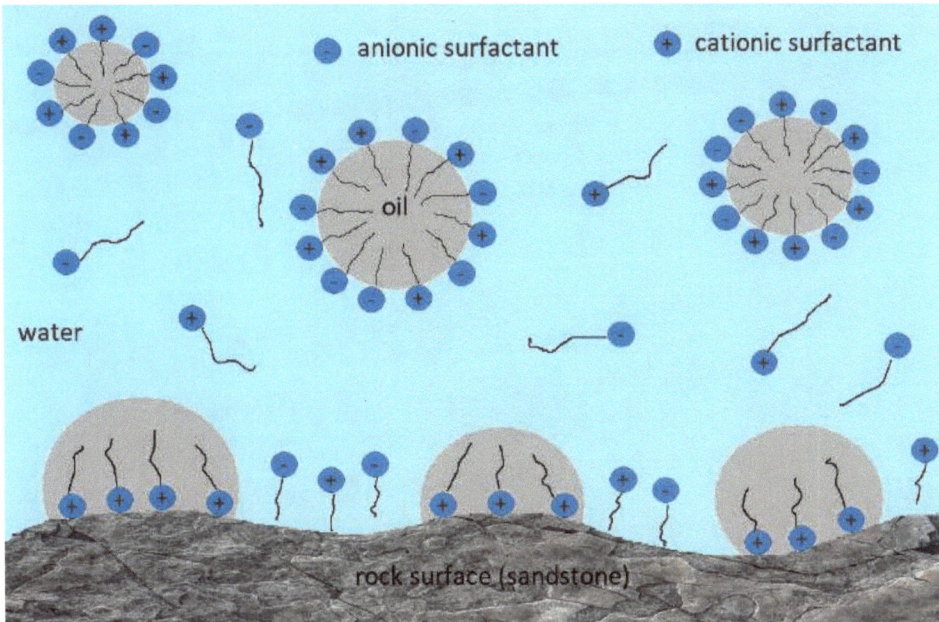

Fig. (8.2). Wettability alteration mechanism by the mixed system consisting of cationic and anionic surfactants.

Researchers from India, analyzing protic ILs derived from trialkylammonium sulfates, acetates and phosphates, observed that the ILs with longer alkyl chain lengths showed much lower IFT than the ILs with shorter alkyl chain lengths for the oil / water system [20], confirming the lowest IFT values for tripropyl-ammonium bisulfate, hydroxypropyl ammonium trifluoroacetate and triethylammonium acetate. At the same time, these very ILs displayed an important effect on the oil recovery with 17-23% of additional oil by injecting them into a porous medium represented by a sand packing, polymeric solution, low-salinity brine, and IL, which was improved even more up to 28% under high-salinity brine conditions. The results of these additional recovery experiments are in good agreement with the IFT studies (Fig. **8.3**) [21].

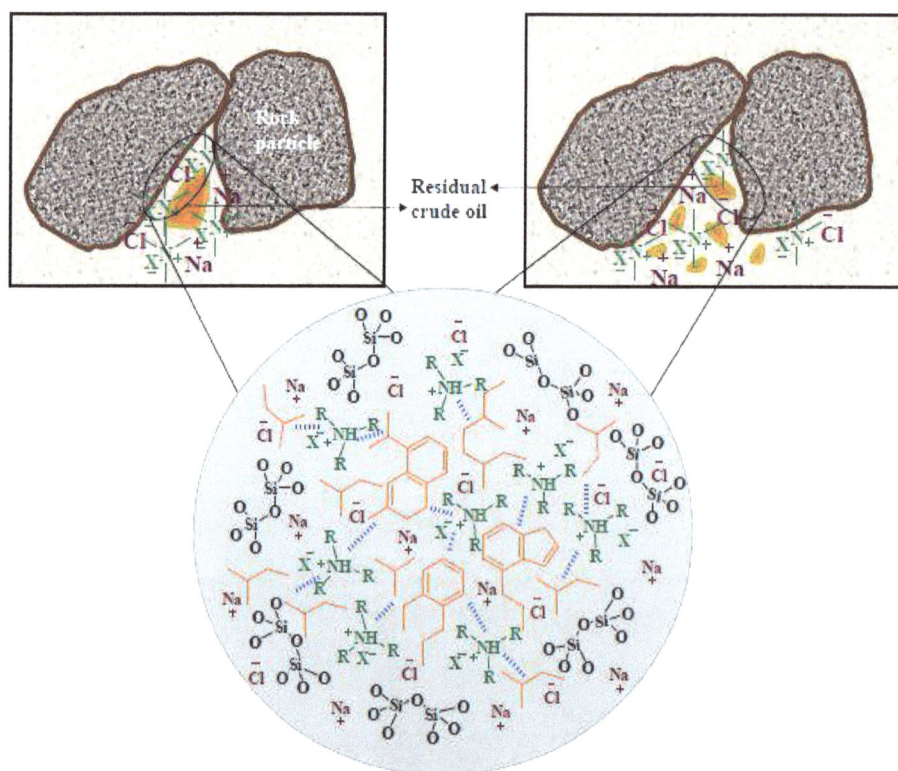

Fig. (8.3). Interaction among ILs, oil and rock in the presence of brine [21].[Reprinted with permission from Sakthivel, S.; Gardas, R. L.; Sangwai, J. S. Effect of Alkyl Ammonium Ionic Liquids on the Interfacial Tension of the Crude Oil–Water System and Their Use for the Enhanced Oil Recovery Using Ionic Liquid-Polymer Flooding. *Energy Fuels* **2016**, *30*, 2514–2523.Copyright 2022 American Chemical Society].

By studying nine ILs, researchers from Saudi Arabia concluded that the IL derived from tetraalkylammonium sulfate showed the best behavior pattern in brine at different temperatures, diminishing the IFT to values up to ≈3 mN/m; at the same time, the IL-based polymers can present the inconvenience of being dissolved in brine, which makes difficult their application in EOR processes [22].

In general, heterocyclic ILs have exhibited higher interfacial activity than aliphatic ILs due to the fact that the hydrophobicity increase in heterocyclic ILs with respect to aliphatic ones could be the reason to higher interfacial activity [23], nevertheless, protic ILs based on butyrolactam and caprolactam (Fig. **8.4**) have presented even lower IFT values than the ILs derived from imidazolium or ammonium [24].

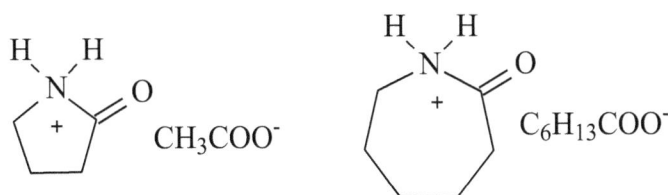

Fig. (8.4). Structure of protic ILs derived from butyrolactam and caprolactam.

The lowest IFT and surface tension values for caprolactam hexanoate are approximately 9.53–8.11 N/m at 1000 ppm (Table **8.1**) [24, 25].

Table 8.1. IFT reduction in the oil/water system in the presence of carboxylate ILs.

ILs	Interfacial Tension (mN/m) at 25 °C and Various IL Concentration		
	0 ppm	50 ppm	1000 ppm
[Butyrolactam]⁺[acetate]⁻	72.0	53.7	52.0
[Butyrolactam]⁺[hexanoate]⁻	72.0	48.9	47.7
[Caprolactam]⁺[formate]⁻	72.0	54.3	52.8
[Caprolactam]⁺[acetate]⁻	72.0	50.2	49.0
[Caprolactam]⁺[hexanoate]⁻	72.0	47.9	46.4
[3-Hydroxypropylammonium]⁺[acetate]⁻	72.0	54.8	53.4
[3-Hydroxypropylammonium]⁺[trifluoroacetate]⁻	72.0	51.1	50.0

There is a series of works related to heavy oil extraction processes (up to 90%) using ILs derived from tetrafluoroborate, trifluoroacetate or trifluoromethanesulfonate (Fig. **8.5**); in addition, ILs with these anions are soluble in water [26 - 29] whereas those derived from hexafluorophosphates and triflates are not soluble in water or in brine in spite of the fact that they are used in EOR processes [30]. Notwithstanding, during the extraction process with ILs featuring organic fluorides in their structure, there is the risk of contaminating the crude oil with this type of halides, which can damage the catalysts for oil hydrodesulfurization processes; for this reason, these compounds have to be removed in spite of the fact that the removal process is complicated with the possibility of reaching a maximal removal percentage of 90% [31, 32].

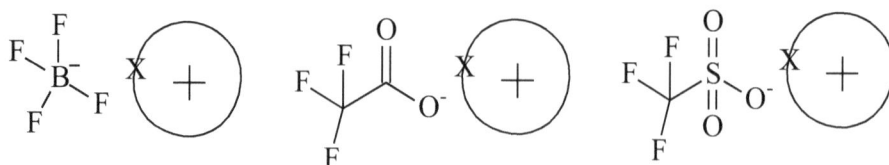

Fig. (8.5). Structure of fluorinated ILs.

Researchers from Malaysia tried to analyze available research works on alternative chemical agents for EOR processes, including ILs [33].

During the last decade, the group led by Dr. E. Buenrostro-González at the IMP carried out several studies and even technological tests based on zwitterionic surfactants (Fig. **8.6**), which played the role of multifunctional agents for removal processes of organic compound deposits, oil recovery, asphaltene dispersion, viscosity reduction of heavy crude oil, rock wettability modification and relative permeability. According to the authors, the action mechanism of the zwitterionic surfactant during the recovery process consists of two stages: oil production and alteration of the rock wettability. The oil production stage is carried out as follows: the interaction between asphaltene and zwitterionic surfactant creates an ion-dipole pair, which captures another asphaltene molecule by means of an ion-dipole pair, forming a supramolecular complex, which in turn also captures an oil molecule with their further liberation. At the rock wettability alteration stage, a zwitterionic surfactant molecule is transported to the rock by the formed ion-dipole pair, where asphaltenes help it pass through the oily phase until reaching the rock surface. The ion-dipole pair is adsorbed on the rock surface and then, the asphaltene is separated from the zwitterionic liquid, being adhered to the rock surface. Consequently, due to the polar characteristics of the ionic molecule, the rock wettability is altered and as a result, the porous medium will be less oil wet and more water wet [34].

The researchers performed theoretical studies based on the density-functional theory (DFT) and provided information on the geometry, stability, electrochemistry, reactivity and selectivity of the zwitterionic molecules; in addition, through the Mulliken population analysis, it was shown that the gemini zwitterionic liquids can form supramolecular structures with other molecules through ion-ion, ion-dipole, dipole-dipole and hydrogen bonds and interact with the surface of calcite rock, changing the wettability of the calcite-water-oil system [35].

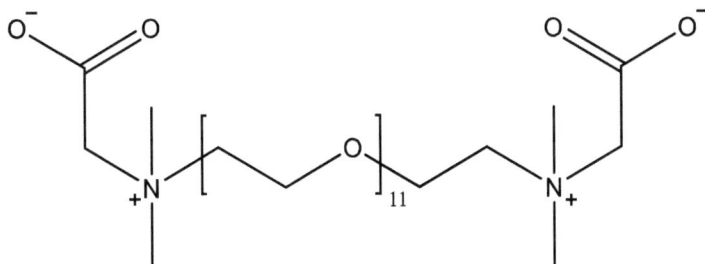

Fig. (8.6). General structure of the Gemini zwitterionic liquid.

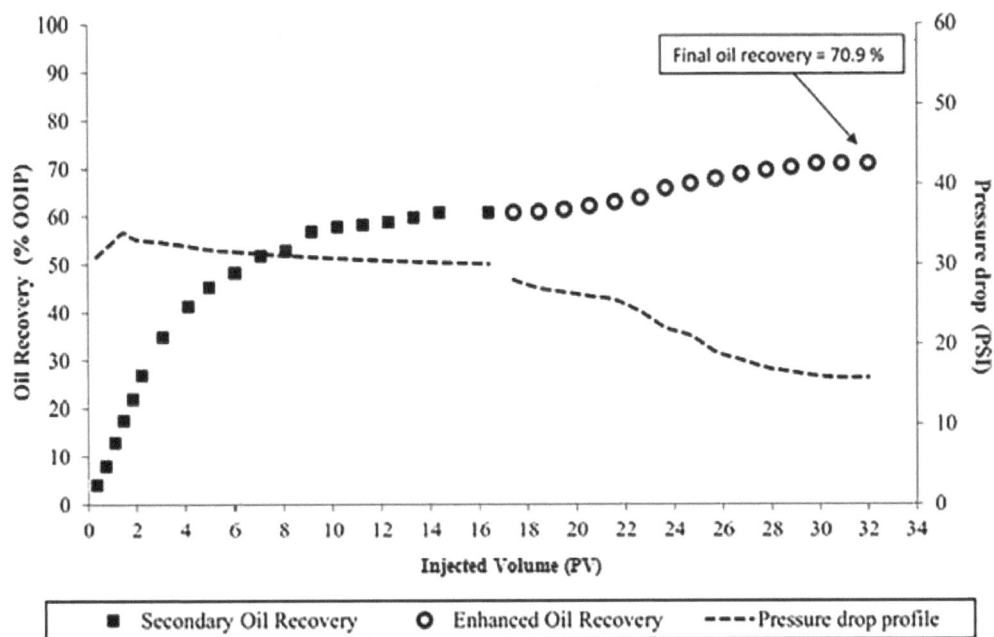

Fig. (8.7). Percentage of accumulated oil and pressure difference during the displacement test employing the zwitterionic surfactant solution [36].[Reprinted with permission from Alcázar-Vara, L. A.; Zamudio-Rivera, L. S.; Buenrostro-González, E. Application of Multifunctional Agents During Enhanced Oil Recovery. In *Chemical Enhanced Oil Recovery (cEOR):* a Practical Overview. IntechOpen, 2016, 10.5772/64792].

The zwitterionic surfactant derived from alkyl betaine at 1 g/L is capable of reducing the IFT in a system consisting of light crude oil (31 °API) and brine (2.6

wt.% of NaCl) up to 0.3 mN/m while at lower concentration (0.1 g/L), the contact angle between oil and the solid dolomite surface changes from 0° to 30°, thus confirming its efficiency as wettability modifier. The asphaltene dispersion efficiency increases as the concentration of the zwitterionic surfactant also augments, however, it is stabilized at 0.5 g/L. In addition, at this concentration, the zwitterionic surfactant diminishes the apparent viscosity of crude oil from 8000 cP to 6500 cP. Through the laboratory EOR tests, an increase of 10% of additional recovery was registered during the injection process of such surfactant at 0.1% (Fig. **8.7**) [36]. The use of ILs and deep eutectic solvents as surfactants for chemical EOR operations has been described in the review by Mert Atilhan and Santiago Aparicio [37].

In the last five years, IL-based emulsions for EOR processes have been developed at the IMP. The emulsions were Winsor I oil-in-water type with dispersed phase drop size below 4 microns. During the laboratory EOR tests using Berea rock cores, the emulsion with 1% of IL was applied as a batch equivalent to 10% of porous volume after the secondary recovery process by water injection, succeeding in getting additional oil recovery up to 28% (Fig. **8.8**). The filtration model was used to describe the emulsion transport through the porous medium; this model is based on the local permeability reduction (trapping) and not on viscosity alteration. The emulsion is trapped and freed from the rock through a non-linear way, thus reducing the permeability of the porous medium [38 - 40].

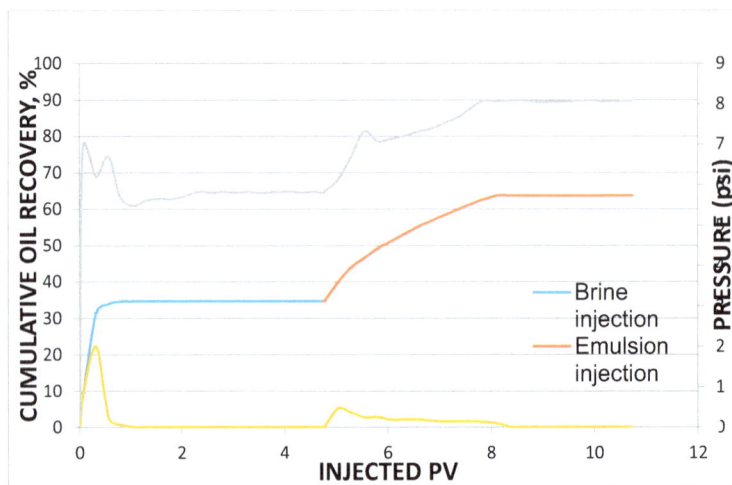

Fig. (8.8). Percentage of cumulative oil and pressure difference during the displacement test using brine and emulsion at 1% within the rock core at 78 °C [39].

Recently, worldwide, the Researchers' attention has been drawn by polymers featuring IL-based monomers in their structure, as such polymers possess both high thermal stability and resistance in salty water medium; furthermore, these compounds can reduce the IFT and are compatible with reservoir rocks and fluids, altering the rock wettability to water wet [41]. But most importantly, these polymers offer the possibility of designing them ad hoc according to the application need. Most EOR projects related to the injection of polymers or alkali-surfactant-polymer (ASP) flooding are devoted to sandstone-type oil fields [5].

Weber *et al.* tried to systematize and understand the relationship among the chemical structure of polymers, their action mechanism in brine and rheological properties for their possible EOR application by offering many examples of associative polymers and polymers with siloxane, anionic and cationic monomers in their structure [42].

In this sense, the copolymer with two monomers, 2-acrylamide-2-methylpropane sulfonate and methyl acrylate, presented the lowest IFT values at 1000 ppm and contact angle change from 105 ° to 60 ° at 1500 ppm; for this reason, the rock wettability alteration is the main mechanism taking place during the recovery process of additional oil; for this purpose, the copolymer mentioned above, at 3000 ppm and diluted in seawater, is capable of increasing the additional oil recovery in about 6% for carbonate medium and 10% for sandstone medium [43]. Over the last decade, the company SNF has developed various ionic polymers based on acrylamide, 2-acrylamido-2-methylpropane sodium sulfonate (ATBS) and vinylpyrrolidone for EOR processes [4, 44, 45]; the terpolymers can be applied in EOR processes considering reservoirs with temperatures up to 120 °C and salinity levels as high as 70 g/L, keeping the initial polymer viscosity for 1 year. However, the use of vinylpyrrolidone increases the polymer cost, but diminishes the terpolymer molecular weight. For this reason, the developments by SNF, over the last 3 years, have been focused on vinylpyrrolidone-free polymers (Table **8.2**) [44, 46] with different ATBS contents, which has allowed a dose reduction of 50%, achieving the same viscosity.

Furthermore, the vinylpyrrolidone-free copolymers displayed better thermal stability, encouraging their application in EOR processes considering reservoirs with temperatures as high as 140 °C and congenital water salinity up to 230 g /L [47]; even in the presence of sulfhydric acid (500 ppm) or 150 ppb of oxygen [48]. At the same time, such polymers can be used in the formulations to produce polymeric gels [49].

Table 8.2. Viscosity data of polymers with and without vinylpyrrolidone in their structure after the aging process at 120 and 140 °C.

Polymer		Residual Viscosity at 7.3 s⁻¹(%)											
		6 Months						12 Months					
		Seawater		Brine (100,000 ppm TDS)		Brine (230,000 ppm TDS)		Seawater		Brine (100,000 ppm TDS)		Brine (230,000 ppm TDS)	
		120°C	140°C	120°C	140°C	120°C	140°C	120°C	140°C	120°C	140°C	120°C	140°C
With NVP	SAV333	67	66	68	68	88	44	60	50	58	54	66	34
Without NVP	SAV28 with low ATBS concentration	62	16	74	14	68	17	54	5	68	7	74	6
	SAV19 with median ATBS concentration	81	18	75	26	95	31	84	4	71	8	76	6
	SAV10 with high ATBS concentration	100	74	99	84	86	92	76	44	85	79	82	63

Strictly speaking, polymers derived from 2-acrylamido-2-methylpropane sodium sulfonate, despite being ionic, are not derived from ILs, as the ionic monomer cation is sodium [50].

Kathmann and McCormick published a cycle of works related to the synthesis of zwitterionic copolymers derived from acrylamide and 4-(2-acrylamid-2-methylpropyldimethylammonium) and their evaluation in electrolytic media, concluding that the solubility of these copolymers depends on the amount of zwitterionic monomer incorporated to these compounds. If the concentration of the ionic monomer is above 25% mol, the copolymers are only swellable in aqueous media; then, the concentration of such monomer should be below 10% mol. In deionized water at pH 8, the polymers displayed high viscosity values due to intermolecular associations, which can be interrupted by adding salts. Notwithstanding, at low pH, all the copolymers exhibited improved viscosity due to charge-charge repulsions between the quaternary ammonium groups [51 - 59].

Lin Ye carried out the synthesis and evaluation of different polymers based on hydroxyethylcellulose modified hydrophobically with a cationic monomer of 4-isopropylbenzyl chloride. The use of a suitable amount of hydrophobic monomer can ensure both strong intermolecular association and good water solubility of the modified polymer, accomplishing high apparent viscosity up to 700 mPa/s at polymer concentration of 0.9 g/dL, even in salty water at 1.5% of NaCl, and surface activity [60]. By including fragments of cationic monomers, such as *N*-alkyldecyl-N, N-dimethylammonium chloride, in the hydroxyethylcellulose structure, the apparent viscosity was improved up to 2000 mPa/s at 1 wt.% (Fig. **8.9**); however, by increasing the temperature, shear rate or medium salinity, the viscosity of polymers diminishes [61].

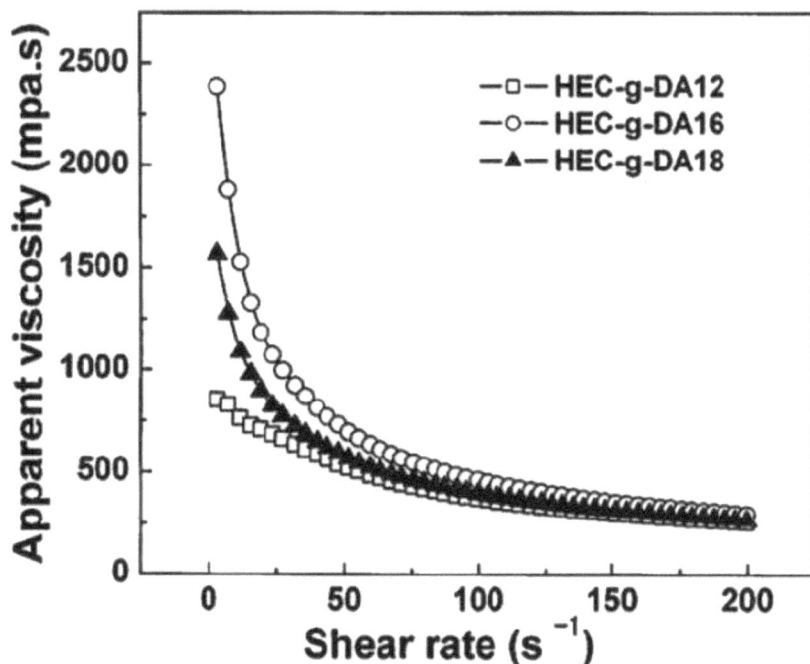

Fig. (8.9). Effect of the shear rate on the apparent viscosity of cationic polymers derived from hydroxyethylcellulose at 1 wt.% [59]. [Reprinted with permission from McCormick, C. L.; Salazar, L. C. Water-Soluble Copolymers. 43. Ampholytic Copolymers of Sodium 2-(Acrylamido)-2-Methylpropanesulfonate with [2-(Acrylamido)-2-Methylpropyl] Trimethylammonium Chloride. *Macromolecules*, **1992**, *25*, 1896–1900. Copyright 2022 American Chemical Society].

The polymer poly[$C_{16}VIm^{+}$][Br^{-}] derived from vinylimidazole at 200 ppm presented IFT reduction up to 1.38 mN/m and thanks to the amphiphilic nature of the IL, the polymer molecules are adsorbed spontaneously on the interface. The core rock experiments showed additional oil recovery as high as 31% after applying a polymer slug equivalent to 0.5 VP [62].

At the IMP, in the last two years, the synthesis and evaluation of ionic terpolymers containing monomers such as acrylamide, vinylpyrrolidone and ILs was carried out in order to use them in EOR processes. By employing the theoretical experiments based on the Fox-Flory equation, a correlation between the gyration radium of the polymers in saline medium and viscosity reduction was established. The results regarding the reduced viscosity were compared with the corresponding experimental ones, thus establishing a weight prediction for the polymer poly[acrylamide-co-vinylpyrrolidone-co-(chloride vinyl benzyl) trimethylammonium] (Fig. **8.10**) of 281130 g/mol, which was compared with the experimental value of 324000g/mol. The deviation in the molecular weight values is due to the polymer-polymer interactions that were not considered in the numerical models of the polymer in brine used in this work [63].

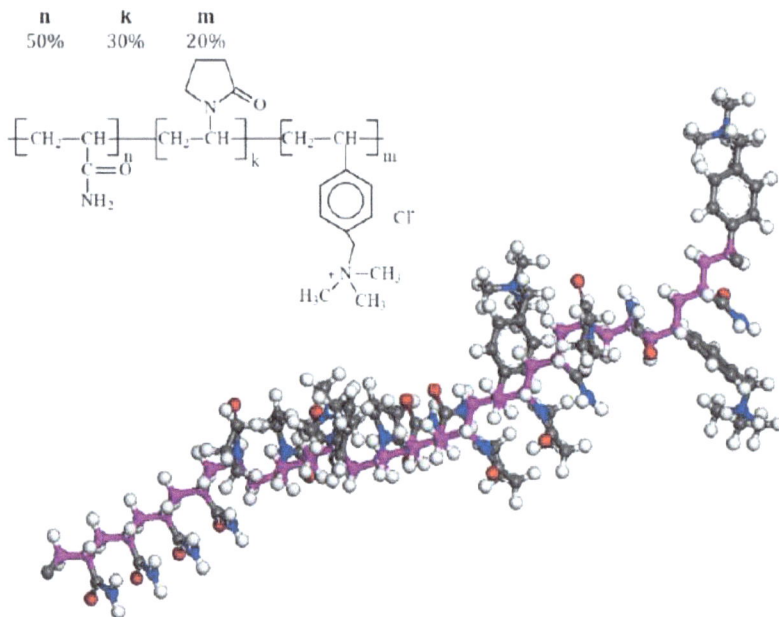

Fig. (8.10). Molecular structure of the terpolymer poly[acrylamide-co-vinylpyrrolidone-co-(chloride vinyl benzyl) trimethylammonium] [63]. [Reprinted with permission from Vega-Paz, A.; Guevara-Rodríguez, F. de J.; Palomeque-Santiago, J. F.; Likhanova, N. V. Polymer Weight Determination from Numerical and Experimental Data of the Reduced Viscosity of Polymer in Brine. *Rev. Mex. Física*, **2019**, *65*, 321–327. Copyright 2022 Revista Mexicana de Física].

In the same way, by means of molecular dynamic simulation, the behavior prediction of terpolymers in brine with monovalent and bivalent metal ions at 160 °C was carried out, finding that the radial distribution function of salt ions such as sodium and calcium around the polymers is related to the polymer resistance to brine. This property was correlated with experimental evidences, which allowed to conclude why the commercial terpolymer type ATBS Super pusher SAV 225, with a negative charge, had a viscosity loss of 57% in 7 days whereas one of the synthesized cationic terpolymers diminished its viscosity in 40% after 28 days of exposure to brine at 160 °C (Fig. **8.11**).

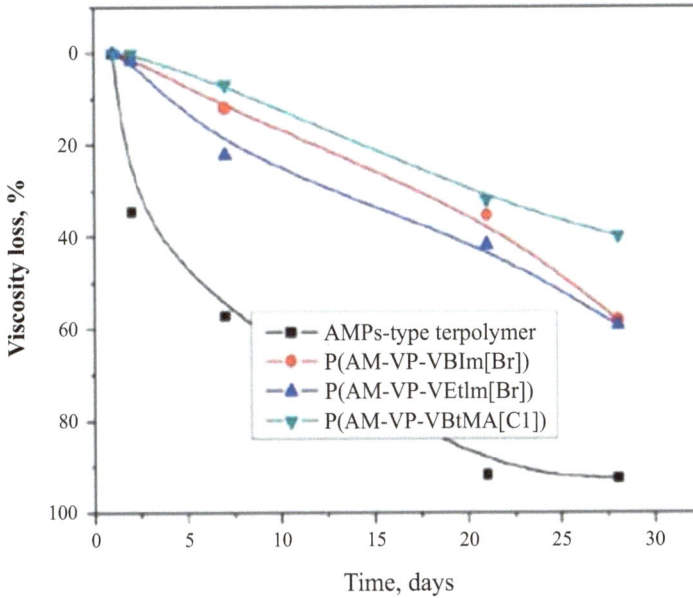

Fig. (8.11). Viscosity loss of the polymeric solution after brine exposure for several days at 160 °C [64]. [Reprinted with permission from Victorovna Likhanova, N.; Guzmán-Lucero, D.; Palomeque-Santiago, J. F.; Guevara-Rodríguez, F. de J. Molecular Dynamics Simulation for Salinity Resistance Prediction of Cationic Terpolymers at High Temperature. *Mol. Phys.***2020**, *118*, 1–12. Copyright 2022 Taylor & Francis Group].

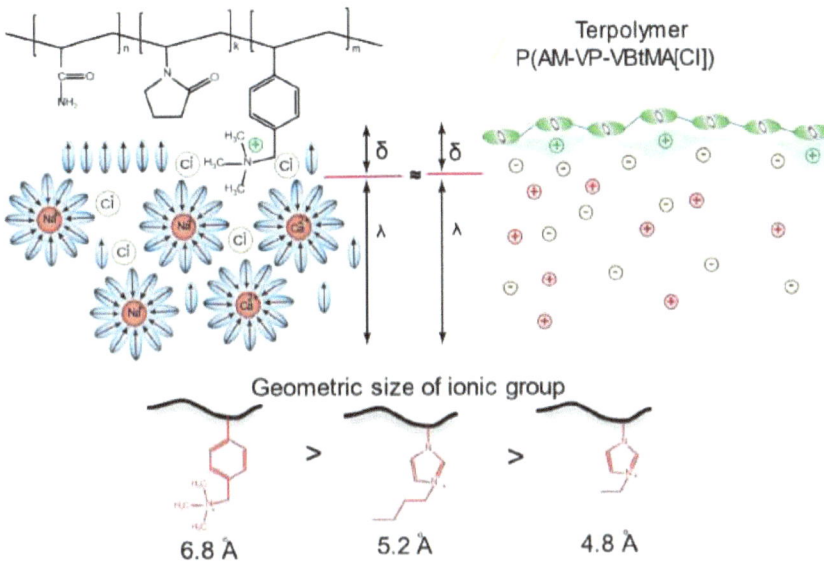

Fig. (8.12). Schematic representation of the electrical double layer around the polymer in brine [64]. [Reprinted with permission from Victorovna Likhanova, N.; Guzmán-Lucero, D.; Palomeque-Santiago, J. F.; Guevara-Rodríguez, F. de J. Molecular Dynamics Simulation for Salinity Resistance Prediction of Cationic Terpolymers at High Temperature. *Mol. Phys.* **2020**, *118*, 1–12. Copyright 2022 Taylor & Francis Group].

In addition, the molecular dynamic simulation indicated that, in general, negatively charged polymers have better water affinity than polymers with global positive charge and by increasing the side alkyl chain length in terpolymers derived from imidazole, the affinity to the aqueous medium is diminished. At the same time, the radial distribution showed that for negatively charged terpolymers, the concentration of calcium and sodium around them is 20-30 times higher than around positively charged polymers, being less resistant to high salt concentration levels in the aqueous medium. The polymer efficiency was kept in the following order: P (AM-VP-VBtMA [Cl]) → P (AM-VP-VBIm [Br]) → P (AM-VP-VEtIm [Br]) → anionic terpolymer. Although the composition and percentage of the ionic unit of the synthesized terpolymers are similar, the best performance displayed by P (AM-VP-VBtMA [Cl]) could be explained by the geometrical size of the ionic groups (Fig. **8.12**) [64].

In spite of the fact that ILs are promising for EOR applications due to their capability to reduce the IFT or modify the rock wettability, their cost is a limiting factor to move from laboratory level studies to oil field applications.

CONCLUDING REMARKS

Until today there are no chemical products or mixture of these that meets all the requirements to give an excellent performance in polymer flooding process in extreme conditions of salinity and temperature, that is, it maintains its properties and especially viscosity for long periods of time (at last one year), at temperature higher than 140°C, and salinities higher than 20,000 ppm showed in some real well. There are some prototypes of emerging polymers that have shown very attractive properties for this application. In this context, ILs play an important role as best performing polymers for this application. One of the more promising products for this purpose is the gemini ILs which function as surfactants and can reduce the IFT up to 99% at concentrations below 0.2 mol/L, forming a stable emulsion with crude oil.

REFERENCES

[1]   Reservas de Hidrocarburos al 1 de Enero de 2015. Petróleos Mexicanos: México, **2015**. (in Spanish) https://www.gob.mx/cms/uploads/attachment/file/460767/Analisis_de_Reservas_1P_2P_3P_2019._vf-cnh-web.pdf

[2]   Petróleos   Mexicanos: México;   2019.   (in   Spanish)   https://www.gob.mx/cms/uploads/ attachment/file/460767/Analisis_de_Reservas_1P_2P_3P_2019._vf-cnh-web.pdf

[3]   *Libro Blanco de Producción de Hidrocarburos Periodo 2012-2018*; Petróleos Mexicanos: México, **2018**. (in Spanish).

[4]   Thomas, A. *Essentials of Polymer Flooding Technique*; John Wiley & Sons Ltd: New Jersey, **2019**. [http://dx.doi.org/10.1002/9781119537632]

[5]   Gbadamosi, A.O.; Junin, R.; Manan, M.A.; Agi, A.; Yusuff, A.S. An overview of chemical enhanced

oil recovery: recent advances and prospects. *Int. Nano Lett.,* **2019**, *9*(3), 171-202.
[http://dx.doi.org/10.1007/s40089-019-0272-8]

[6]   Hou, B.; Wang, Y.; Cao, X.; Zhang, J.; Song, X.; Ding, M.; Chen, W. Surfactant-Induced Wettability Alteration of Oil-Wet Sandstone Surface: Mechanisms and Its Effect on Oil Recovery. *J. Surfactants Deterg.,* **2016**, *19*(2), 315-324.
[http://dx.doi.org/10.1007/s11743-015-1770-y]

[7]   Pan, F.; Zhang, Z.; Zhang, X.; Davarpanah, A. Impact of anionic and cationic surfactants interfacial tension on the oil recovery enhancement. *Powder Technol.,* **2020**, *373*, 93-98.
[http://dx.doi.org/10.1016/j.powtec.2020.06.033]

[8]   Jia, H.; Lian, P.; Leng, X.; Han, Y.; Wang, Q.; Jia, K.; Niu, X.; Guo, M.; Yan, H.; Lv, K. Mechanism studies on the application of the mixed cationic/anionic surfactant systems to enhance oil recovery. *Fuel,* **2019**, *258*, 116156.
[http://dx.doi.org/10.1016/j.fuel.2019.116156]

[9]   Nasirpour, N.; Mohammadpourfard, M.; Zeinali Heris, S. Ionic liquids: Promising compounds for sustainable chemical processes and applications. *Chem. Eng. Res. Des.,* **2020**, *160*, 264-300.
[http://dx.doi.org/10.1016/j.cherd.2020.06.006]

[10]   Mohammed, M.; Babadagli, T. Wettability alteration: A comprehensive review of materials/methods and testing the selected ones on heavy-oil containing oil-wet systems. *Adv. Colloid Interface Sci.,* **2015**, *220*, 54-77.
[http://dx.doi.org/10.1016/j.cis.2015.02.006] [PMID: 25798909]

[11]   Sakthivel, S.; Velusamy, S.; Gardas, R.L.; Sangwai, J.S. Use of Aromatic Ionic Liquids in the Reduction of Surface Phenomena of Crude Oil–Water System and their Synergism with Brine. *Ind. Eng. Chem. Res.,* **2015**, *54*(3), 968-978.
[http://dx.doi.org/10.1021/ie504331k]

[12]   Manshad, A.K.; Rezaei, M.; Moradi, S.; Nowrouzi, I.; Mohammadi, A.H. Wettability alteration and interfacial tension (IFT) reduction in enhanced oil recovery (EOR) process by ionic liquid flooding. *J. Mol. Liq.,* **2017**, *248*, 153-162.
[http://dx.doi.org/10.1016/j.molliq.2017.10.009]

[13]   Saien, J.; Asadabadi, S. Salting out effects on adsorption and micellization of three imidazolium-based ionic liquids at liquid–liquid interface. *Colloids Surf. A Physicochem. Eng. Asp.,* **2014**, *444*, 138-143.
[http://dx.doi.org/10.1016/j.colsurfa.2013.12.060]

[14]   Sakthivel, S.; Velusamy, S.; Gardas, R.L.; Sangwai, J.S. Nature Friendly Application of Ionic Liquids for Dissolution Enhancement of Heavy Crude Oil, SPE-178418-MS. In: *SPE Annual Technical Conference and Exhibition*; Society of Petroleum Engineers: Texas, **2015**; pp. 28-30.
[http://dx.doi.org/10.2118/178418-MS]

[15]   Yahya, M.S.; Sangapalaarachchi, D.M.T.; Lau, E.V. Effects of carbon chain length of imidazolium-based ionic liquid in the interactions between heavy crude oil and sand particles for enhanced oil recovery. *J. Mol. Liq.,* **2019**, *274*, 285-292.
[http://dx.doi.org/10.1016/j.molliq.2018.10.147]

[16]   Jiao, J.; Han, B.; Lin, M.; Cheng, N.; Yu, L.; Liu, M. Salt-free catanionic surface active ionic liquids 1-alkyl-3-methylimidazolium alkylsulfate: Aggregation behavior in aqueous solution. *J. Colloid Interface Sci.,* **2013**, *412*, 24-30.
[http://dx.doi.org/10.1016/j.jcis.2013.09.001] [PMID: 24144370]

[17]   Pillai, P.; Mandal, A. Wettability Modification and Adsorption Characteristics of Imidazole-Based Ionic Liquid on Carbonate Rock: Implications for Enhanced Oil Recovery. *Energy Fuels,* **2019**, *33*(2), 727-738.
[http://dx.doi.org/10.1021/acs.energyfuels.8b03376]

[18]   Kharazi, M.; Saien, J.; Yarie, M.; Zolfigol, M.A. The superior effects of a long chain gemini ionic liquid on the interfacial tension, emulsification and oil displacement of crude oil-water. *J. Petrol. Sci.*

Eng., **2020**, *195*, 107543.
[http://dx.doi.org/10.1016/j.petrol.2020.107543]

[19]   Gou, S.; Yin, T.; Yan, L.; Guo, Q. Water-soluble complexes of hydrophobically modified polymer and surface active imidazolium-based ionic liquids for enhancing oil recovery. *Colloids Surf. A Physicochem. Eng. Asp.*, **2015**, *471*, 45-53.
[http://dx.doi.org/10.1016/j.colsurfa.2015.02.022]

[20]   Sakthivel, S.; Velusamy, S.; Gardas, R.L.; Sangwai, J.S. Adsorption of aliphatic ionic liquids at low waxy crude oil–water interfaces and the effect of brine. *Colloids Surf. A Physicochem. Eng. Asp.*, **2015**, *468*, 62-75.
[http://dx.doi.org/10.1016/j.colsurfa.2014.12.010]

[21]   Sakthivel, S.; Gardas, R.L.; Sangwai, J.S. Effect of Alkyl Ammonium Ionic Liquids on the Interfacial Tension of the Crude Oil–Water System and Their Use for the Enhanced Oil Recovery Using Ionic Liquid-Polymer Flooding. *Energy Fuels*, **2016**, *30*(3), 2514-2523.
[http://dx.doi.org/10.1021/acs.energyfuels.5b03014]

[22]   Bin Dahbag, M.; AlQuraishi, A.; Benzagouta, M. Efficiency of ionic liquids for chemical enhanced oil recovery. *J. Pet. Explor. Prod. Technol.*, **2015**, *5*(4), 353-361.
[http://dx.doi.org/10.1007/s13202-014-0147-5]

[23]   Nandwani, S.K.; Malek, N.I.; Chakraborty, M.; Gupta, S. Insight into the Application of Surface-Active Ionic Liquids in Surfactant Based Enhanced Oil Recovery Processes–A Guide Leading to Research Advances. *Energy Fuels*, **2020**, *34*(6), 6544-6557.
[http://dx.doi.org/10.1021/acs.energyfuels.0c00343]

[24]   Velusamy, S.; Sakthivel, S.; Gardas, R.L.; Sangwai, J.S. Substantial Enhancement of Heavy Crude Oil Dissolution in Low Waxy Crude Oil in the Presence of Ionic Liquid. *Ind. Eng. Chem. Res.*, **2015**, *54*(33), 7999-8009.
[http://dx.doi.org/10.1021/acs.iecr.5b01337]

[25]   Sakthivel, S.; Chhotaray, P.K.; Velusamy, S.; Gardas, R.L.; Sangwai, J.S. Synergistic effect of lactam, ammonium and hydroxyl ammonium based ionic liquids with and without NaCl on the surface phenomena of crude oil/water system. *Fluid Phase Equilib.*, **2015**, *398*, 80-97.
[http://dx.doi.org/10.1016/j.fluid.2015.04.011]

[26]   Sui, H.; Zhang, J.; Yuan, Y.; He, L.; Bai, Y.; Li, X. Role of binary solvent and ionic liquid in bitumen recovery from oil sands. *Can. J. Chem. Eng.*, **2016**, *94*(6), 1191-1196.
[http://dx.doi.org/10.1002/cjce.22477]

[27]   Cao, N.; Almojtaba Mohammed, M.; Babadagli, T. Wettability Alteration of Heavy-Oil-Bitum-n-Containing Carbonates by Use of Solvents, High-pH Solutions, and Nano/Ionic Liquids. *SPE Reservoir Eval. Eng.*, **2017**, *20*(2), 363-371.
[http://dx.doi.org/10.2118/183646-PA]

[28]   Mohammed, M.A.; Babadagli, T. Experimental Investigation of Wettability Alteration in Oil-Wet Reservoirs Containing Heavy Oil. *SPE Reservoir Eval. Eng.*, **2016**, *19*(4), 633-644.
[http://dx.doi.org/10.2118/170034-PA]

[29]   Pillai, P.; Kumar, A.; Mandal, A. Mechanistic studies of enhanced oil recovery by imidazolium-based ionic liquids as novel surfactants. *J. Ind. Eng. Chem.*, **2018**, *63*, 262-274.
[http://dx.doi.org/10.1016/j.jiec.2018.02.024]

[30]   Pereira, J.F.B.; Costa, R.; Foios, N.; Coutinho, J.A.P. Ionic liquid enhanced oil recovery in sand-pack columns. *Fuel*, **2014**, *134*, 196-200.
[http://dx.doi.org/10.1016/j.fuel.2014.05.055]

[31]   Bera, A.; Belhaj, H. Ionic liquids as alternatives of surfactants in enhanced oil recovery—A state-o--the-art review. *J. Mol. Liq.*, **2016**, *224*, 177-188.
[http://dx.doi.org/10.1016/j.molliq.2016.09.105]

[32]   Babío Núñez, B. Estudio de Procesos de Hydrocracking y Mild Hydrocracking: Evaluación de Catalizadores. In: *Planta Piloto y Desarrollo de Modelos Para El Proceso Industrial*; Universidad Politécnica de Madrid: España, **2017**.

[33]   Tackie-Otoo, B.N.; Ayoub Mohammed, M.A.; Yekeen, N.; Negash, B.M. Alternative chemical agents for alkalis, surfactants and polymers for enhanced oil recovery: Research trend and prospects. *J. Petrol. Sci. Eng.,* **2020**, *187*, 106828.
[http://dx.doi.org/10.1016/j.petrol.2019.106828]

[34]   Alcázar-Vara, L.A.; Zamudio-Rivera, L.S.; Buenrostro-González, E.; Hernández-Altamirano, R.; Mena-Cervantes, V.Y.; Ramírez-Pérez, J.F. Multifunctional Properties of Zwitterionic Liquids. Application in Enhanced Oil Recovery and Asphaltene Aggregation Phenomena. *Ind. Eng. Chem. Res.,* **2015**, *54*(11), 2868-2878.
[http://dx.doi.org/10.1021/ie504837h]

[35]   López-Chavez, E.; Garcia-Quiroz, A.; Gonzalez-Garcia, G.; Orozco-Duran, G.E.; Zamudio-Rivera, L.S.; Martinez-Magadan, J.M.; Buenrostro-Gonzalez, E.; Hernandez-Altamirano, R. Quantum chemical characterization of zwitterionic structures: Supramolecular complexes for modifying the wettability of oil–water–limestone system. *J. Mol. Graph. Model.,* **2014**, *51*, 128-136.
[http://dx.doi.org/10.1016/j.jmgm.2014.04.013] [PMID: 24907932]

[36]   Alcázar-Vara, L.A.; Zamudio-Rivera, L.S.; Buenrostro-González, E. Application of Multifunctional Agents During Enhanced Oil Recovery. In: *Chemical Enhanced Oil Recovery (cEOR): a Practical Overview*; IntechOpen, **2016**.
[http://dx.doi.org/10.5772/64792]

[37]   Atilhan, M.; Aparicio, S. Review on chemical enhanced oil recovery: Utilization of ionic liquids and deep eutectic solvents. *J. Petrol. Sci. Eng.,* **2021**, *205*, 108746.
[http://dx.doi.org/10.1016/j.petrol.2021.108746]

[38]   Hernandez-Perez, J.R.; Likhanova, N.V.; Lopez-Falcon, D.A.; Guzman-Lucero, D.; Olivares-Xometl, O.; Garcia-Rodriguez, S. Simulation of Emulsion Flooding in Porous Media. *Pan American Mature Fields Congress,* Veracruz, México, **2015**.

[39]   Likhanova, N.V.; Demikhova, I.I.; Hernandez Perez, J.R.; Moctezuma Berthier, A.E.; Olivares Xometl, O.; Cuapantecatl Mendieta, M. A. Hydrophobic Compound Emulsions Free of Silicon and Fluorine for an Oil Recovering Method That Modifies the Wettability of Rocks from Hydrophilic to Oleophilic. *US10329474B2,* **2019**.

[40]   Demikhova, I.I.; Likhanova, N.V.; Hernandez Perez, J.R.; Falcon, D.A.L.; Olivares-Xometl, O.; Moctezuma Berthier, A.E.; Lijanova, I.V. Emulsion flooding for enhanced oil recovery: Filtration model and numerical simulation. *J. Petrol. Sci. Eng.,* **2016**, *143*, 235-244.
[http://dx.doi.org/10.1016/j.petrol.2016.02.018]

[41]   Druetta, P.; Raffa, P.; Picchioni, F. Chemical enhanced oil recovery and the role of chemical product design. *Appl. Energy,* **2019**, *252*, 113480.
[http://dx.doi.org/10.1016/j.apenergy.2019.113480]

[42]   Wever, D.A.Z.; Picchioni, F.; Broekhuis, A.A. Polymers for enhanced oil recovery: A paradigm for structure–property relationship in aqueous solution. *Prog. Polym. Sci.,* **2011**, *36*(11), 1558-1628.
[http://dx.doi.org/10.1016/j.progpolymsci.2011.05.006]

[43]   Alhussinan, S.N.; Alyami, H.Q.; Alqahtani, N.B.; AlQuraishi, A.A. Chemical Enhanced Oil Recovery With Water Soluble Poly Ionic Liquids in Carbonate Reservoirs, SPE-192373-MS. *SPE Kingdom of Saudi Arabia Annual Technical Symposium and Exhibition; Society of Petroleum Engineers; Saudi Arabia,* **2018**.
[http://dx.doi.org/10.2118/192373-MS]

[44]   Gaillard, N.; Giovannetti, B.; Favero, C.; Caritey, J-P.; Dupuis, G.; Zaitoun, A. New Water Soluble Anionic NVP Acrylamide Terpolymers for Use in Harch EOR Conditions, SPE-169108-MS. In: *SPE Improved Oil Recovery Symposium*; Society of Petroleum Engineers: Oklahoma, USA, **2014**; pp. 12-

16.
[http://dx.doi.org/10.2118/169108-MS]

[45]   Guzmán-Lucero, D.; Martínez-Palou, R.; Palomeque-Santiago, J.F.; Vega-Paz, A.; Guzmán-Pantoja, J.; López-Falcón, D.A.; Guevara-Rodríguez, F.J.; García-Muñoz, N.A.; Castillo-Acosta, S.; Likhanova, N.V. Water Control with Gels Based on Synthetic Polymers under Extreme Conditions in Oil Wells. *Chem. Eng. Technol.,* **2022**, *45*(6), 998-1016.
[http://dx.doi.org/10.1002/ceat.202100648]

[46]   Dupuis, G.; Antignard, S.; Giovannetti, B.; Gaillard, N.; Jouenne, S.; Bourdarot, G.; Morel, D.; Zaitoun, A. A New Thermally Stable Synthetic Polymer for Harsh Conditions of Middle East Reservoirs. Part I. Thermal Stability and Injection in Carbonate Cores, SPE-188479-MS. In: *Abu Dhabi International Petroleum Exhibition & Conference*; Society of Petroleum Engineers: Abu Dhabi, UAE, **2017**; pp. 13-16.
[http://dx.doi.org/10.2118/188479-MS]

[47]   Rodriguez, L.; Antignard, S.; Giovannetti, B.; Dupuis, G.; Gaillard, N.; Jouenne, S.; Bourdarot, G.; Morel, D.; Zaitoun, A.; Grassl, B. A New Thermally Stable Synthetic Polymer for Harsh Conditions of Middle East Reservoirs; Society of Exploration Geophysicists. In: *Research and Development Petroleum Conference and Exhibition*; Abu Dhabi, UAE, **2018**.
[http://dx.doi.org/10.2118/190200-MS]

[48]   Masalmeh, S.; AlSumaiti, A.; Gaillard, N.; Daguerre, F.; Skauge, T.; Skuage, A. Extending Polymer Flooding Towards High-Temperature and High-SalinityCarbonate Reservoirs, SPE-197647-MS. In: *Abu Dhabi International Petroleum Exhibition & Conference*; Society of Petroleum Engineers: Abu Dhabi, UAE, **2019**.
[http://dx.doi.org/10.2118/197647-MS]

[49]   Juárez, J.L.; Rodriguez, M.R.; Montes, J.; Trujillo, F.D.; Monzòn, J.; Dupuis, G.; Gaillard, N. Conformance Gel Design for High Temperature Reservoirs, SPE-200640-MS. *SPE Europec featured at 82nd EAGE Conference and Exhibition; Society of Petroleum Engineers; Amsterdam, Netherlands,* **2020**.
[http://dx.doi.org/10.2118/200640-MS]

[50]   Saxena, N.; Kumar, S.; Mandal, A. Adsorption characteristics and kinetics of synthesized anionic surfactant and polymeric surfactant on sand surface for application in enhanced oil recovery. *Asia-Pac. J. Chem. Eng.,* **2018**, *13*(4), e2211.
[http://dx.doi.org/10.1002/apj.2211]

[51]   Kathmann, E.E.; White, L.A.; McCormick, C.L. Water soluble polymers: 69. pH and electrolyte responsive copolymers of acrylamide and the zwitterionic monomer 4-(2-acrylamid--2-methylpropyldimethylammonio) butanoate: synthesis and solution behaviour. *Polymer (Guildf.),* **1997**, *38*(4), 871-878.
[http://dx.doi.org/10.1016/S0032-3861(96)00586-1]

[52]   Kathmann, E.E.; White, L.A.; McCormick, C.L. Water soluble polymers: 70. Effects of methylene *versus* propylene spacers in the pH and electrolyte responsiveness of zwitterionic copolymers incorporating carboxybetaine monomers. *Polymer (Guildf.),* **1997**, *38*(4), 879-886.
[http://dx.doi.org/10.1016/S0032-3861(96)00587-3]

[53]   Kathmann, E.E.; McCormick, C.L. Water-soluble polymers. 72. synthesis and solution behavior of responsive copolymers of acrylamide and the zwitterionic monomer 6-(2-acrylamid--2-methylpropyldimethylammonio) hexanoate. *J. Polym. Sci. A Polym. Chem.,* **1997**, *35*(2), 243-253.
[http://dx.doi.org/10.1002/(SICI)1099-0518(19970130)35:2<243::AID-POLA6>3.0.CO;2-T]

[54]   Kathmann, E.E.L.; Davis, D.D.; McCormick, C.L. Water-Soluble Polymers. 60. Synthesis and Solution Behavior of Terpolymers of Acrylic Acid, Acrylamide, and the Zwitterionic Monomer 3-[(--Acrylamido-2-methylpropyl)dimethylammonio]-1-propanesulfonate. *Macromolecules,* **1994**, *27*(12), 3156-3161.
[http://dx.doi.org/10.1021/ma00090a007]

[55]   Kathmann, E.E.; McCormick, C.L. Water-soluble polymers. 71. pH responsive behavior of terpolymers of sodium acrylate, acrylamide, and the zwitterionic monomer 4-(2-acrylamid--2-methylpropanedimethylammonio)butanoate. *J. Polym. Sci. A Polym. Chem.,* **1997**, *35*(2), 231-242.
[http://dx.doi.org/10.1002/(SICI)1099-0518(19970130)35:2<231::AID-POLA5>3.0.CO;2-V]

[56]   Mccormick, C.; Johnson, C.B. Water-Soluble Polymers. Xxxiv. Ampholytic Terpolymers of Sodium 3-Acrylamido-3-Methylbutanoate with 2-Acrylamido-2-Methylpropane-Dimethylammonium Chloride and Acrylamide: Synthesis and Absorbency Behavior. *J. Macromol. Sci. Part A Pure Appl. Chem.,* **1990**, *27*(5), 539-547.
[http://dx.doi.org/10.1080/10601329008544789]

[57]   McCormick, C.L.; Johnson, C.B. Water-soluble copolymers. 29. Ampholytic copolymers of sodium 2-acrylamido-2-methylpropanesulfonate   with   (2-acrylamido-2-methylpropyl)dimethylammonium chloride: solution properties. *Macromolecules,* **1988**, *21*(3), 694-699.
[http://dx.doi.org/10.1021/ma00181a026]

[58]   McCormick, C.L.; Johnson, C.B. Water-soluble polymers. 28. Ampholytic copolymers of sodium 2-acrylamido-2-methylpropanesulfonate   with   (2-acrylamido-2-methylpropyl)dimethylammonium chloride: synthesis and characterization. *Macromolecules,* **1988**, *21*(3), 686-693.
[http://dx.doi.org/10.1021/ma00181a025]

[59]   McCormick, C.L.; Salazar, L.C. Water-soluble copolymers. 43. Ampholytic copolymers of sodium 2-(acrylamido)-2-methylpropanesulfonate with [2-(acrylamido)-2-methylpropyl]trimethylammonium chloride. *Macromolecules,* **1992**, *25*(7), 1896-1900.
[http://dx.doi.org/10.1021/ma00033a009]

[60]   Dai, S.; Ye, L.; Huang, R. A study on the solution behavior of IPBC-hydrophobically-modified hydroxyethyl cellulose. *J. Appl. Polym. Sci.,* **2006**, *100*(4), 2824-2831.
[http://dx.doi.org/10.1002/app.23743]

[61]   Wang, K.; Ye, L. Solution Behavior of Hydrophobic Cationic Hydroxyethyl Cellulose. *J. Macromol. Sci. Part B Phys.,* **2014**, *53*(1), 149-161.
[http://dx.doi.org/10.1080/00222348.2013.808512]

[62]   Pillai, P.; Mandal, A. A comprehensive micro scale study of poly-ionic liquid for application in enhanced oil recovery: Synthesis, characterization and evaluation of physicochemical properties. *J. Mol. Liq.,* **2020**, *302*, 112553.
[http://dx.doi.org/10.1016/j.molliq.2020.112553]

[63]   Vega Paz, A.; Guevara Rodríguez, F.J.; Palomeque Santiago, J.F.; Victorovna Likhanova, A.N. Polymer weight determination from numerical and experimental data of the reduced viscosity of polymer in brine. *Rev. Mex. Fis.,* **2019**, *65*(4 Jul-Aug), 321-327.
[http://dx.doi.org/10.31349/RevMexFis.65.321]

[64]   Victorovna Likhanova, N.; Guzmán-Lucero, D.; Palomeque-Santiago, J.F.; Guevara-Rodríguez, F.J. Molecular dynamics simulation for salinity resistance prediction of cationic terpolymers at high temperature. *Mol. Phys.,* **2020**, *118*(15), e1718225.
[http://dx.doi.org/10.1080/00268976.2020.1718225]

CHAPTER 9

Applications of ILs as Catalysts in the Reaction to Obtain Alkylate Gasoline

Abstract: Alkylate gasoline is one of the most valuable products produced in the crude oil refining process. This product is one of the main contributors to the gasoline pool because it has a high-octane rating and a low content of contaminants such as sulfur, nitrogen, and aromatic compounds. This chapter reviews the advances in the use of ILs as catalysts for the isobutene/butene reaction with the objective of developing a less hazardous and more sustainable process to obtain alkylate gasoline.

Keywords: Alkylate, Brønsted acids, Butene, High-octane, Ionic liquids, Isobutane, Lewis acids, Lewis/ Brønsted ILs, Low sulfur, Pool, Selectivity, Supported-ILs, Trimethylpentane.

INTRODUCTION

Today, the production of high-quality, environmental-friendly gasoline remains a major challenge. Current and future global specifications require the reduction of aromatics, sulfur and oxygenates in gasoline as a final product. For this reason, gasoline as an end product is obtained by blending hydrocarbons obtained in a series of processes (gasoline pool), including the alkylation reaction. Nowadays, the produced alkylate represents approximately 10% of global gasoline production.

In the oil field, the alkylation reaction between olefins and isobutane is of great interest because it allows the synthesis of 8-carbon synthetic hydrocarbons, mainly trimethylpentanes (TMPs, Fig. **9.1**) and other branched kerosenes, which give rise to gasoline of excellent quality and octane rating, high calorific value, low Reid pressure and low pollutant content, which is known as alkylation gasoline and is a major contributor to the gasoline pool.

This reaction results in a fuel that burns efficiently, increasing the engine life and creating lower emissions. By increasing the octane rating of fuels, the addition of lead to the alkylate is not necessary, thus avoiding the use of an environmentally harmful substance without affecting the engine efficiency [1].

Rafael Martínez Palou & Natalya V. Likhanova

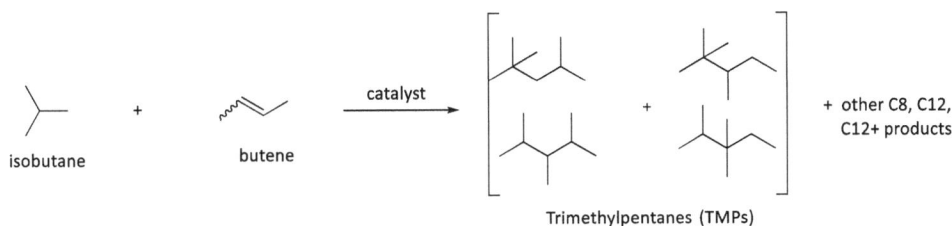

Fig. (9.1). Reaction to obtain alkylate gasoline.

In this context, Phillips alkylation technology is the most widely used in Mexico and internationally, employing HF as a catalyst. HF, besides being a very toxic product, is highly corrosive and must be used with a minimal purity of 99.5% and is used in conjunction with aluminum trichloride, which acts as a water absorbent and favors the catalyst performance [2].

Although in the alkylation process there are controls to retain the fluorinated compounds and HF produced in the process, latent risks cannot be totally averted due to HF leaks in industrial facilities, where clouds with lethal HF doses can form; so, their handling must be extremely careful, and also alkylation gasoline can be contaminated by HF and fluorinated organic compounds that are also highly toxic [3].

In order to contribute to the solution of this drawback, the IMP developed a technology to remove fluorinated compounds from alkylation gasoline, which was discussed in Chapter 2 of this book [4].

The alkylation process with H_2SO_4 is also widely used worldwide as it is a potentially less hazardous technology, although the process is less efficient than with HF, requiring a large amount of catalyst and the octane yield in the alkylate is lower. It also involves a high consumption of water that must be treated to remove the remaining hydrocarbons. The sulfuric acid alkylation technology has certain advantages over H_2SO_4: it is less volatile and toxic, the required isobutane/butene ratio is lower than that of HF, and distillation and isobutane recovery costs are relatively lower [5 - 7].

Because of all the limitations described for the catalysts used commercially in this technology, researchers in the area have devoted great efforts to the development of new, more efficient and environmentally friendly catalysts with the aim of making the process of obtaining alkylation gasoline a more efficient and sustainable technology [8].

In this sense, the alkylation reaction has been extensively studied with solid catalysts such as zeolites [9 - 11], heteropolyacids [12, 13], and Nafion-based nanocomposites [14 - 16]. These types of catalysts have also proved to be very efficient carrying out this reaction, reducing significantly toxicity risks and operating costs by forming a process with a solid heterogeneous catalyst that is neither toxic nor corrosive; however, these technologies have not been extended so far because the catalysts deactivate quickly and present difficulties for regeneration and reuse and tend to form oligomers and unsaturated hydrocarbons [17]. Currently, China has installed a commercial plant with this technology (ALkyclean™), using the zeolite-based catalyst Alkystar™ by Albemarle [18].

ILs as Catalysts to Obtain Alkylate Gasoline

Currently, ILs are the most explored alternative as catalysts for the isobutane/butene alkylation reaction and in our opinion, the most promising candidates to replace the acids used commercially in the alkylation process. Fig. (9.2) shows the different types of ILs with acidic properties that have been studied and those that will be discussed in the following sections.

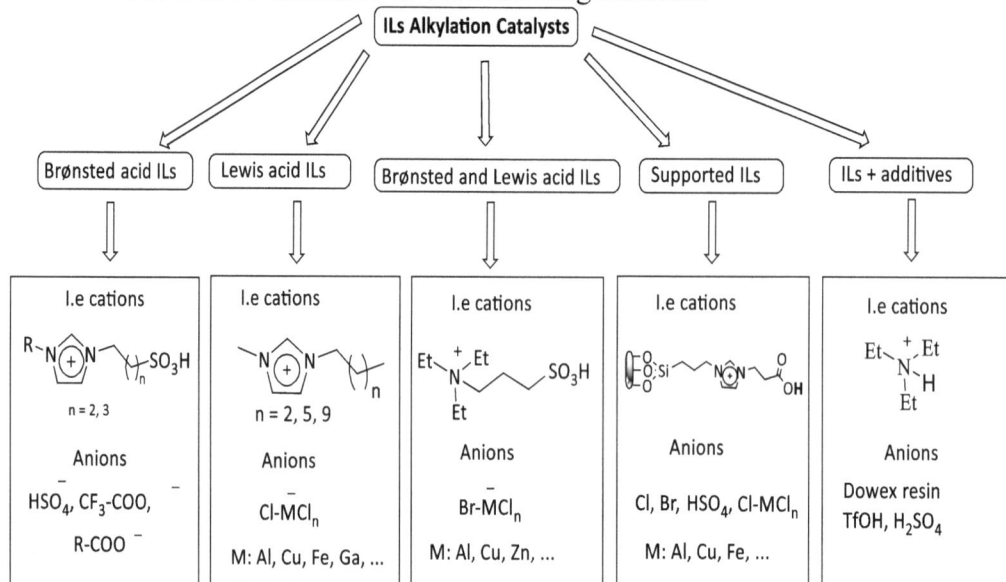

Fig. (9.2). Acidic ILs studied as alkylation reaction catalysts.

This topic has been studied by different authors in recent years [19 - 22], including the review published by IMP researchers titled New Insights into the Progress on the Isobutane/Butene Alkylation Reaction and Related Processes for High-Quality Fuel Production. A Critical Review. Energy Fuels 2020, 34 (12), 15525–15556 [23] by Díaz Velázquez, H.; Likhanova, N.; Aljammal, N.;

Verpoort, F.; and Martínez-Palou, R., which not only discusses recent studies in the development of novel alkylation reaction catalysts based on ILs, but also all the research efforts that have been made with different types of catalysts in order to produce alkylation gasoline through a sustainable way and more efficiently. This paper discusses the mechanism of the alkylation reaction and comments critically on the current limitations and actions for IL-catalyzed alkylation technology to be employed worldwide as one of the most widespread and successful IL-based technologies. Some recent advances in the subject are then discussed.

IMP researchers have also submitted the review titled Recent progress on catalyst technologies for high quality gasoline production by Díaz Velázquez, H, Cerón-Camacho, R, M. Mosqueira, L. L., Hernández-Cortez, J. G., Montoya de la Fuente, J. A., Hernández-Pichardo, M. L., and Martínez-Palou, R. This paper review summarized the more recent advances in the development of catalysts for the different processes used by the refining industry to improve the composition and properties of gasoline fuels and reduce the content of pollutants such as sulfur, nitrogen and aromatic compounds. This work includes the critical review of processes such as catalytic cracking, hydrocracking, reforming, isomerization and alkylation. Additionally, a section is dedicated to additives for gasoline production and advances in the development of catalysts for Fischer-Tropsch process are also discussed.

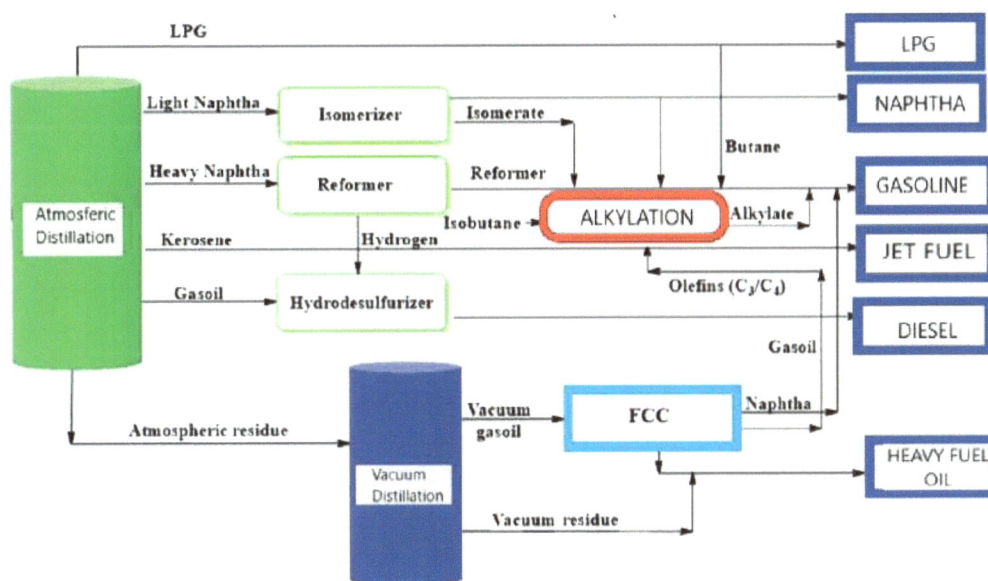

Fig. (9.3). Simplified scheme of an FCC refinery process.

Fig. (**9.3**) shows a simplified scheme of an FCC refinery with the different processes that contribute to the main end products and those that have to do with the final composition of gasoline, highlighting in red the alkylation process discussed in this chapter, which is one of the main contributors to the production of high-octane, low-pollutant gasoline.

Brønsted Acid ILs as Catalysts of Isobutane/Butene Alkylation

Acidic Brønsted ILs refer to those ILs that contain acidic protons in the cation, anion or both. These compounds have attractive properties such as high chemical stability, tunable acidity, high catalytic activity, and better dissolving capacity for a wide range of organic and inorganic compounds [24, 25]. These compounds have been widely explored as catalysts in organic synthesis [26], and the isobutane/butene alkylation reaction has been no exception, for example, Sun *et al.* studied a series of HSO_3-functionalized imidazolium-based ILs, mixed with H_2SO_4 as synergistic catalysts for isobutane alkylation. MD simulation and DFT calculations have also been employed to further correlate them with the corresponding catalytic performance. It has been observed that these pure ILs have relatively low acidity (Ho = 1.52 - 1.62) compared to H_2SO_4 (Ho = 12.0). The acidity value decreases with the increasing chain length of the alkyl group linked to the imidazolium chain (Fig. **9.4**).

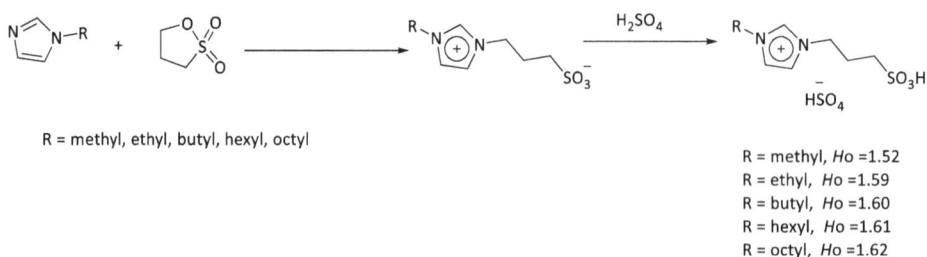

R = methyl, ethyl, butyl, hexyl, octyl

R = methyl, Ho =1.52
R = ethyl, Ho =1.59
R = butyl, Ho =1.60
R = hexyl, Ho =1.61
R = octyl, Ho =1.62

Fig. (9.4). Synthesis and acidity value of Brønsted acidic ILs.

This acidity range is low to achieve high conversion of the alkylation reaction, in particular to obtain TMPs. On the other hand, it is widely known that acidity plays an important role in the alkylate quality [27]. It has been observed that the mixture of these Brønsted acid catalysts form a synergistic mixture with H_2SO_4, which improves the efficiency and performance of the catalyst. The catalyst efficiency is improved as the imidazolium alkyl chain grows longer, which appears to be related to the higher solubility of isobutane, which has been further correlated with the nanostructured-aggregation feature of the longer alkyl chain, which has also been confirmed by MD simulation. In addition, the maximal catalytic efficiency has been achieved with 10% of catalyst. On the other hand, the [OPSIm][HSO$_4$]/H$_2$SO$_4$ mixture has contributed to the C8 selectivity up to 75.73 and RON up to 95.66. The improved catalytic performance can be attributed to the enhanced isobutane dispersion resulting from the SFILs with longer alkyl chain. The reusability of the SFILs/H$_2$SO$_4$ mixture can reach up to 22 runs, outclassing the pure H$_2$SO$_4$.

Another alternative to Brønsted acidic ILs is protic ILs in which the proton is located in the IL cation. An example of these catalysts was studied by Wang *et al.* who prepared protic ILs of the [N$_{222}$H]$^-$-type by reacting triethylamine with different acid ratios of trifluoromethanesulfonic acids and H$_2$SO$_4$. The [N$_{222}$H]CF$_3$SO$_3$/TfOH compound with χ_{TfOH} = 0.69 was the prototype that showed the best performance and could be reused for at least 36 times without a significant loss of catalytic activity, which was attributed to the buffering effect of the anionic groups, [N$_{222}$H][CF$_3$SO$_3$(CF$_3$SO$_3$H)x] (x = 0, 1, 2); the selectivity to C8 was of 86.2% and the RON of 97.3. The new catalytic system, in addition to its reusability, had the advantage of showing better compatibility with impurities and was significantly less corrosive for carbon and stainless steel types than pure sulfuric and pure triflic acids [28]. The authors proposed that the acidity could be buffered by binding the solubilized acid in the anionic cluster form, thus [CF$_3$SO$_3$(CF$_3$SO$_3$H)x] with (x = 0, 1, 2) as it has been proposed by other authors [29].

Similar results regarding the synergy produced by mixing Brønsted acidic ILs with strong acids have been obtained in other recent works [30, 31].

In 2017, Zhao *et al.* studied an alternative to improve the distribution of alkylation reaction products through mass transfer enhancement by the addition of ILs with surfactant properties such as 1-dodecyl-N,N,N-trimethylammonium hydrogen sulfate ([N$_{12,1,1,1}$)][HSO$_4$]. The enhanced product distribution and alkylate quality were attributed to the good dispersion of the hydrocarbon in the acid, increasing the hydrocarbon solubility in the catalytic system, *i.e.* enhancing the mass transfer in the reaction system. In conclusion, adding a quaternary ammonium surfactant is a promising method to improve the quality of the alkylate and decrease the acid consumption in the sulfuric acid catalyzed C4 alkylation, which would benefit the environment [32].

Recently, it has been found that trifluoromethanesulfonic acid and taurine form a deep eutectic solvent in polyethylene glycol that shows excellent performance as a catalyst for the alkylation reaction with conversion and selectivity to C8 of 82.5 and 85.5, respectively. Moreover, [TfOH]3[TAU]/PEG-200 exhibited high catalytic activity and excellent recyclability under a low alkane/olefin ratio, providing a simple, economic, highly efficient, and green route for the preparation of higher octane alkylate gasoline [33].

Lewis Acid ILs as Catalysts of Isobutane/Butene Alkylation

Lewis acid ILs refer to those types of compounds in which the anion contains a tetrahalometalate-type anion and that are generally obtained through a metathesis reaction of a halogenated anion with a transition metal anion in which the salt must be in a molar ratio greater than 0.5 with respect to the IL precursor to confer Lewis acidic properties [34, 35]. The Lewis acidity exhibited by these ILs will depend on the type of metal and molar ratio between the precursor halogenated IL and metal salt [19].

The molar fraction of the Lewis acid MCl_n can be calculated according to Equation 1:

$$X\,MCl_3 = \text{mol of } MCl_n / (\text{mol of } MCl_n + \text{mol of halogenated IL precursor}) \quad \textbf{(1)}$$

Some examples of Lewis acids that have been studied as catalysts for the alkylation reaction are shown in Fig. (**9.5**).

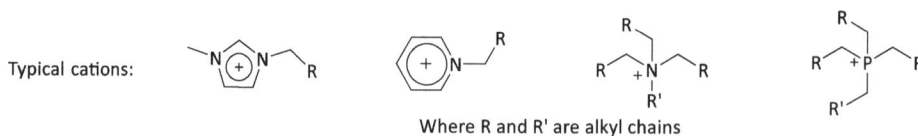

Typical cations:

Where R and R' are alkyl chains

Anions: X/MX$_n$ (typical ratio: 0.51-0.79), X = Cl, Br, and M are metals as Al, Fe, Ga, Zn, Cu, Ga

Fig. (9.5). Typical structures of cations and anions of Lewis acidic ILs studied as catalysts.

The most studied ones are those containing the anion based on haloaluminates, which present a great catalytic performance in the conversion and selectivity of the isobutane/butene reaction under "mild temperature and pressure" conditions and with the use of a ratio that is lower than that of the charge employed with conventional acid catalysts.

For this reason, ILs belonging to this family have been extensively studied as catalysts for the alkylation reaction. Table **9.1** summarizes some of the most important articles related to the study of Lewis acids used as catalysts of the alkylation reaction.

Table 9.1. Summary of recent articles on the use of Lewis acidic ILs as catalysts to produce alkylate gasoline.

IL Structure	Reaction Conditions	Main Results	Observations	Ref.
$Al_2Cl_6Br^-$ $Al_2Cl_7^-$	[OMIM]Br/AlCl3 as catalysts (content of water in HCs= 30 ppm; (−5 °C; 600 kPa; P/O ratio (mol)= 13; stirring time= 60 min; IL/olefin mol ratio= 0.4; molar fraction of $AlCl_3$= 0.6). $Et_3NHCl/AlCl_3$ IL catalysts. (−5 °C; 600 kPa; P/O ratio (mol)= 13; stirring time= 15 min; IL/olefin mol ratio= 0.4; molar fraction of $AlCl_3$ = 0.6. Some additives such as water, acidic cationic exchange resin and transition metal salts were used.	High content of the desired trimethylpentanes (TMPs) (up to 72.3%) and thus a high research octane number (RON up to 98) in the alkylate were obtained by using CuCl modified triethylamine hydrochloride aluminum chloride ([Et_3NH]Cl/$AlCl_3$) (molar fraction of $AlCl_3$ = 0.6) at −5 °C for 15 min.	Using transition metal salts raised the selectivity of ILs and thus the alkylate quality, especially when $Et_3NHCl/AlCl_3$ was used. Metal salts increased the acidity of [OMIM]Br/$AlCl_3$, but mostly the Lewis acidity; therefore, the alkylate quality was improved, but not so much.	[36]
$Cl/AlCl_3^-$ $(X_{AlCl3}= 0.64)$	T = 5°C, P = 6 bar, stirring: 1800 rpm, P/O ratio=100.	The effect of the *in-situ* addition of anhydrous hydrogen chloride was studied. Addition of HCl reactivated an already deactivated IL catalyst that showed decreased activity and selectivity to TMPs, the primarily desired high-octane compounds of alkylation. To prevent the complete deactivation of the IL catalyst, the biphasic system was saturated with HCl gas before starting the alkylation, to give constant activity and excellent selectivity of TMPs.	The solubility of HCl in the IL was higher than the organic feed phase and increased with a decrease in temperature. To verify data found in the literature that suggest a relationship between the occurrence of conjunct polymers and CAIL deactivation process, conjunct polymers were extracted from an IL catalyst and their mass fraction was determined by GC analysis.	[37]

(Table 9.1) cont.....

IL Structure	Reaction Conditions	Main Results	Observations	Ref.
  SbF_6^- n = 2, 4, 6 SbF_6^- n = 0, 2, 4, 6 ILs were mixed with CF_3SO_3H ILs were mixed with CF_3SO_3H	Volume of hydrocarbon feed (molar I/O = 10) = 50 mL, volume of catalyst= 25 mL, P= 0.5 MPa, T= 10 °C, stirring= 1000 r/min, reaction time= 10 min.	The optimal alkylate with C8 selectivity of up to 80% and RON of 95 was obtained using the acidic IL [HMIm] [SbF_6] (47.9 wt.%)/CF_3SO_3H (52.1 wt.%) as catalyst. The results were better than those obtained by using either neat H_2SO_4 or CF_3SO_3H as catalyst.	The coupled IL/acid system could be easily adjusted by varying the cation/anion combination or by modifying the structures of the cation or anion. RON between 87 and 95 were obtained with all the ILs containing [SbF_6]$^-$ as anion.	[38]
Basic IL Cl/AlCl$_3$ Wet and dry composite ILs Cl/AlCl$_3$/CuCl	A molar ratio of isobutane/butenes of 20 was introduced into the reactor at 5 mL/min flow rate. Stirring rate: 500 rpm, Temperature controlled by a thermostatic oil bath at 23°C.	Lewis acidic species were mainly provided by Al_2Cl_7, represented by a broad signal at 102 ppm in the ^{27}Al NMR spectrum. Lewis acidic species were demonstrated by using pyridine as an indicator giving a pyridine-Lewis acid complex. The Brønsted acidic species were related to Al_2Cl_6OH (97 ppm) in the ^{27}Al NMR spectrum. This anion could lose HCl, *e.g.* by evacuation to form non-reactive chlorooxyaluminate (Al_2Cl_5O) at 94 ppm.	A study of the species present with different molar ratios of metal salts by IR, ^1H and ^{27}Al NMR was carried out.	[39]

(Table 9.1) cont.....

IL Structure	Reaction Conditions	Main Results	Observations	Ref.
Cl/AlCl₃/CuCl X_{AlCl3}=0.65-1.0 Cl/AlCl₃/CuCl	Catalyst, 150 mL; feedstock, 400 mL; feeding rate, 750 mL/h; reaction temperature, 15°C; stirrer speed, 1400 r/min; and I/O molar ratio, 20:1.	Under the same amide/AlCl₃ molar ratio, a NMA-AlCl₃-based IL analogue showed better catalytic performance than other amide-AlCl₃-based IL analogues and Et₃NHCl-AlCl₃ IL. This finding was mainly attributed to the NMA-AlCl₃-based IL. When the structure of the 0.75NMA-1.0AlCl₃-based IL analogue was modified by CuCl, the selectivity to C8 significantly increased from 76.18 to 94.65 wt.%, in which the TMPs/DMHs molar ratio and alkylate RON reached 14.98 and 98.40, respectively.	The effects of the amide structure, amide/ AlCl₃ molar ratio, and CuCl modification on the catalytic performance were investigated in semicontinuous mode. ²⁷Al NMR spectra further confirmed that a new composite anion [AlCl4CuCl] was formed in the CuCl-modified NMA-AlCl3 IL analogue, which effectively inhibited the side reactions and resulted in better C8 selectivity	[40]

Very recently, Wu *et al.* carried out a new study on the effect of adding aromatics on improving the performance of alkylation catalysts of the Lewis acidic IL-type ([BMIIM]Cl/AlCl₃). Interaction between aromatics (benzene, toluene, xylene and chlorobenzene) and anions was investigated by *in-situ* infrared spectroscopy, Raman spectroscopy and nuclear magnetic resonance which revealed the presence of new species of aromatic–AlCl₄⁻ and [aromatic–H]⁺. ²⁷Al NMR showed that the addition of aromatics prolonged the lifetime of Al species, which indicated interaction between aromatics and ions. The presence of new ions modified the acidic environment of the alkylation reaction to some extent and inhibited the formation of C_{5-7} and C_{9+} by-products. Molecular dynamics (MD) simulation was used to investigate the concentration of isobutane and aromatics at the acid–hydrocarbon interface.

These new species buffer the acidity at a lower level and modify the activity of CILs. Theoretical calculations showed that only the aromatics, enriched at the acid–hydrocarbon interface and with appropriate charge transfer, can effectively affect the CIL alkylation. The aromatic–ion interaction inhibited the polymerization side reactions and promoted hydride transfer from isobutane to C_8^+ [41].

Because of their catalytic performance, chloroaluminate-based ILs are currently used as catalysts in the process known commercially as IsoAlky, which was developed by Chevron and marketed by Honeywell UOP [42].

However, chloroaluminate ILs display high moisture sensitivity, so they generally decompose after the first catalytic cycle in the presence of feeds with water contents higher than 10 ppm [43], and then, hydrogen chloride is injected for *in-situ* regeneration of the catalyst [44]. For this reason, research for the development of new IL-based catalysts with greater chemical stability goes on in order to develop an efficient and sustainable process that can displace the commercial catalysts currently in use.

As described in some examples in Table **9.1**, within the family of Lewis acidic ILs are also considered ILs known as composites (CILs), which contain a complex anion of the halometalate-type formed by two metals [45 - 47].

In 2015, IL catalyzed isobutane/2-butene alkylation modified with metal compounds was studied by Liu *et al*. They found that adding a small amount of transition metal compound could not significantly change the acid strength of the IL, but could increase the I/O ratio of the reaction system. For example, the $CuAlCl_4$ anion could help increase the I/O ratio due to the Cu(I)-butene complexation. The trace amounts of Al-transition metal species like $CuAlCl_4$ and $CuAlBrCl_3$ were detected in the IL catalysts, which adsorbed 2-butene in the alkylation process. The IL $[Et_3NH]Cl-AlCl_3/CuAlCl_4$ produced alkylate with 87.5 wt.% [48].

In the last decade, several patents protecting the use of Lewis acid ILs and composite-type catalysts have been submitted [49 - 51].

China University of Petroleum has developed an alkylation process based on composite ILs known as Ionikylation that is completely non-corrosive, allowing to construct all process equipment with carbon steel. This technology is licensed from Well Resources Inc [52 - 54]. This company has installed several plants with different capacities and this year, it has announced the start-up of a new plant in Jiujiang City capable of producing over 300,000 ton/year of alkylate [55].

Brønsted-Lewis acid ILs

In 2009, Bui *et al*. showed that when a mixture of $[OMIM]Br/AlCl_3$ and $[(HO_3SBu)MIM]HSO_4$ was employed as a catalyst of the alkylation reaction, where a synergistic effect was observed instead of using the independent components. Similar results were obtained for the catalytic performance of $[Et_3NH]Cl/AlCl_3$ *versus* $[OMIM]Br/AlCl_3$ due to Brønsted acidity of the first one as confirmed by FT-IR. Moreover, the Lewis acid significantly improved the chemical stability in the presence of water, so that the catalytic system could be recycled and reused for several reaction cycles without significant loss of its catalytic activity [56].

Liu *et al.* published two papers employing ILs containing Brønsted and Lewis sites in the same structure, which showed excellent catalytic performance and chemical stability compared to the previously described Lewis or Brønsted acidity-only catalysts. In their first work, a family of dual ILs was synthesized from the alkylation of triethylamine with propylsultone and the generation of several dual catalysts by carrying out the metathesis reaction with different metal salts (Fig. **9.6**).

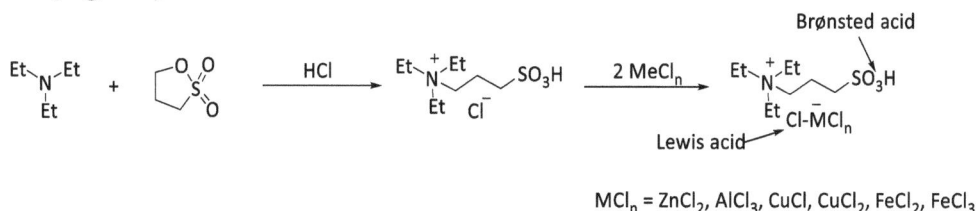

$MCl_n = ZnCl_2, AlCl_3, CuCl, CuCl_2, FeCl_2, FeCl_3$

Fig. (9.6). Synthesis of dual ILs studied by Liu *et al.* in 2015.

The catalyst $[HO_3SC_3NEt_3]Cl\text{-}ZnCl_2$ (molar fraction of $ZnCl_2$ (x), 0.83) was the most effective one and could be reused in ten reaction cycles without noticeable loss of catalytic effectiveness; in all cases, conversions higher than 99% and more than 88% of TMP content were obtained. The reusability of the IL phase was attributed to the alkyl sulfonic acid group covalently tethered to the IL cation and to the Lewis acidic center of the IL, which is inert and stable to water or the alkyl sulfonic acid [57].

Similar results were obtained by Liu *et al.* in 2018 when dual ILs from 1-methylimidazole, like 1-(3-sulfonic acid)-propyl-3-methylimidazolium chlorozincate $[HO_3S\text{-}(CH_2)_3\text{-}MIM]Cl\text{-}ZnCl_2$ (x = 0.67), where synthesized and evaluated as catalysts of the isobutane/butene reaction. The results were explained by the appropriate double acidic site balance and the promoting effect of water on the formation and transfer of protons.

$[HO_3S\text{-}(CH_2)_3\text{-}MIM]Cl\text{-}ZnCl_2$ was reused in seven reaction cycles maintaining a conversion of 100% and selectivity to TMPs between 77-81%. According to the authors, the high effectiveness of this catalyst was due to the formation of the dioctahedral complex through the reaction of the $Zn_2Cl_5\text{-}$ anion with water, whereby the formation of the carbocation and proton transfer between the carbocation and isobutene can be accelerated. (Fig. **9.7**) [58].

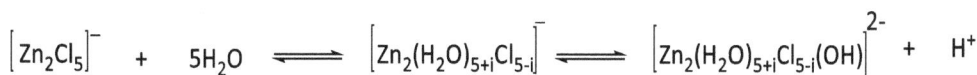

$$\left[Zn_2Cl_5\right]^- + 5H_2O \rightleftharpoons \left[Zn_2(H_2O)_{5+i}Cl_{5-i}\right]^- \rightleftharpoons \left[Zn_2(H_2O)_{5+i}Cl_{5-i}(OH)\right]^{2-} + H^+$$

Fig. (9.7). Formation of the dioctahedral complex from the reaction between the anion $Zn_2Cl_5^-$ and water.

In our opinion, these results have set a precedent that should be further developed because these catalysts have shown the best performance and chemical stability results that could surpass those obtained with prototypes featuring only Lewis or Brønsted acidic catalytic sites.

Supported-ILs

Other IL-based catalysts that have been tested in the isobutane/butene alkylation reaction are ILs supported on inorganic materials such as silica and the like. The IL can be physically supported by impregnation on the support or by anchoring it through the formation of a covalent bond between the IL and support (Fig. **9.8**).

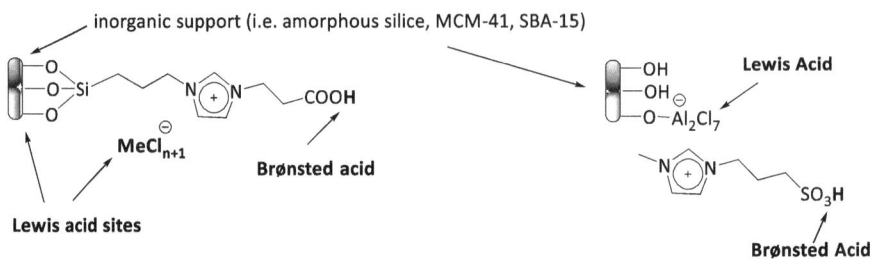

Fig. (9.8). Examples of supported acid ILs containing Lewis and Brønsted acid sites.

Although so far this strategy has been scarcely explored as an alternative to obtain alkylation gasoline, the results described so far are very attractive [15, 59, 60]. These materials allow the development of heterogeneous solid-liquid catalysis in which it is possible to modulate the acidity of the catalyst and gain the advantages of heterogenic catalysis, such as the decrease of the required IL amount, increasing the chemical stability of the catalyst and facilitating its regeneration by avoiding cross contamination and separation of the IL from the hydrocarbon. An example of these results was published in 2020 by Liu *et al*. These authors employed a rotating packed bed reactor to enhance the mass transfer to carry out the alkylation reaction using supported-ILs. Under optimum conditions (isobutane-to-olefins = 7:1; volume ratio of acid-to-hydrocarbon = 1.0; temperature = 5 °C, and rotational speed = 1200 rpm), the TMP content and RON of alkylates were 87.1 wt.% and 99.8, respectively. A large interfacial area was beneficial to the increase in the yield and quality of alkylates [61].

The IMP is currently carrying out a project in which the use of poly(ionic liquids)s (poly(IL)s) and supported-ILs with more than one Brønsted and Lewis acid sites as catalysts of the alkylation reaction is more advantageous than the employment of commonly inorganic acids, as these compounds feature low toxicity, volatility and corrosivity. On the other hand, these supported ILs or forming polymers have an advantage over traditional ILs and especially over chloroaluminate-based ILs, because they can be designed with different ILs anchored to the support or forming a polymer in such a way that they can contain different Brønsted and Lewis acid sites, which allows the design of a catalyst with adequate acidity and stability to achieve an excellent catalytic performance in the alkylation reaction.

CONCLUDING REMARKS

As we have seen in the present chapters, ILs with acidic properties have proved to be the alternative with the best prospects to replace the technologies currently used at the industrial level to obtain alkylation gasoline based on HF and H_2SO_4. Currently, there are technological developments in ILs based on Lewis acids, such as those based on halo aluminate anions and composites, by the addition of CuCl to halo aluminate-containing IL, *i.e.* $Et_3NHCl1.5AlCl_3$, that require a continuity in research in the search for prototypes with better performance, selectivity, and chemical stability than the current prototypes. Supported-ILs and Brønsted and Lewis acidic ILs have proven to be excellent alternatives to catalyze the isobutane/butene alkylation reaction due to their high performance and chemical stability; however, research on these types of ILs is still scarce and needs to be deepened.

REFERENCES

[1]   Hommeltoft, S.I. Isobutane alkylation. *Appl. Catal. A Gen.,* **2001**, *221*(1-2), 421-428.
       [http://dx.doi.org/10.1016/S0926-860X(01)00817-1]

[2]   Linn, C.B.; Grosse, A.V. Alkylation of Isoparaffins by Olefins in Presence of Hydrogen Fluoride. *Ind. Eng. Chem.,* **1945**, *37*(10), 924-929.
       [http://dx.doi.org/10.1021/ie50430a012]

[3]   Weitkamp, J.; Traa, Y. Isobutane/butene alkylation on solid catalysts. Where do we stand? *Catal. Today,* **1999**, *49*(1-3), 193-199.
       [http://dx.doi.org/10.1016/S0920-5861(98)00424-6]

[4]   Miranda, A.D.; Gallo, M.; Domínguez, J.M.; Sánchez-Badillo, J.; Martínez-Palou, R. Experimental and theoretical assessment of the interactions of ionic liquids (ILs) with fluoridated compounds (HF, R-F) in organic medium. *J. Mol. Liq.,* **2019**, *276*, 779-793.
       [http://dx.doi.org/10.1016/j.molliq.2018.12.040]

[5]   Albright, L.F. Alkylation of Isobutane with C_3 –C_5 Olefins To Produce High-Quality Gasolines: Physicochemical Sequence of Events. *Ind. Eng. Chem. Res.,* **2003**, *42*(19), 4283-4289.
       [http://dx.doi.org/10.1021/ie0303294]

[6]   Xin, Y.; Hu, Y.; Li, M.; Chi, K.; Zhang, S.; Gao, F.; Jiang, S.; Wang, Y.; Ren, C.; Li, G. Isobutane Alkylation Catalyzed by H_2SO_4: Effect of H_2SO_4 Acid Impurities on Alkylate Distribution. *Energy Fuels,* **2021**, *35*(2), 1664-1676.
       [http://dx.doi.org/10.1021/acs.energyfuels.0c03453]

[7]   Furimsky, E. Spent refinery catalysts: Environment, safety and utilization. *Catal. Today,* **1996**, *30*(4), 223-286.

[http://dx.doi.org/10.1016/0920-5861(96)00094-6]

[8]   Singh, S.K.; Savoy, A.W. Ionic liquids synthesis and applications: An overview. *J. Mol. Liq.,* **2020,** *297,* 112038.
[http://dx.doi.org/10.1016/j.molliq.2019.112038]

[9]   Primo, A.; Garcia, H. Zeolites as catalysts in oil refining. *Chem. Soc. Rev.,* **2014,** *43*(22), 7548-7561.
[http://dx.doi.org/10.1039/C3CS60394F] [PMID: 24671148]

[10]   Vogt, E.T.C.; Whiting, G.T.; Dutta Chowdhury, A.; Weckhuysen, B.M. Zeolites and Zeotypes for Oil and Gas Conversion. In: *Advances in Catalysis*; Elsevier, **2015**; 58, pp. 143-314.
[http://dx.doi.org/10.1016/bs.acat.2015.10.001]

[11]   Zeolites and Related Microporous Materials. https://www.elsevier.com/books/zeolites-and-relat-d-microporous-materials-state-of-the-art-1994/holderich/978-0-444-81847-8

[12]   Baronetti, G.; Thomas, H.; Querini, C.A. Wells–Dawson heteropolyacid supported on silica: isobutane alkylation with C4 olefins. *Appl. Catal. A Gen.,* **2001,** *217*(1-2), 131-141.
[http://dx.doi.org/10.1016/S0926-860X(01)00576-2]

[13]   Sarsani, V.R.; Wang, Y.; Subramaniam, B. Toward Stable Solid Acid Catalysts for 1-Butene + Isobutane Alkylation: Investigations of Heteropolyacids in Dense CO $_2$ Media. *Ind. Eng. Chem. Res.,* **2005,** *44*(16), 6491-6495.
[http://dx.doi.org/10.1021/ie048911v]

[14]   Rorvik, T.; Dahl, I.M.; Mostad, H.B.; Ellestad, O.H. Nafion-H as Catalyst for Isobutane/2-Butene Alkylation Compared with a Cerium Exchanged Y Zeolite. *Catal. Lett.,* **1995,** *33,* 127-134.
[http://dx.doi.org/10.1007/BF00817052]

[15]   Shen, W.; Gu, Y.; Xu, H.; Dubé, D.; Kaliaguine, S. Alkylation of isobutane/1-butene on methyl-modified Nafion/SBA-15 materials. *Appl. Catal. A Gen.,* **2010,** *377*(1-2), 1-8.
[http://dx.doi.org/10.1016/j.apcata.2009.12.012]

[16]   Botella, P.; Corma, A.; López-Nieto, J.M. The Influence of Textural and Compositional Characteristics of Nafion/Silica Composites on Isobutane/2-Butene Alkylation. *J. Catal.,* **1999,** *185*(2), 371-377.
[http://dx.doi.org/10.1006/jcat.1999.2502]

[17]   Guisnet, M.; Ribeiro, F.R. *Deactivation and Regeneration of Zeolite Catalysts*; World Scientific, **2011**.
[http://dx.doi.org/10.1142/p747]

[18]   AlkyClean® | Lummus Technology. https://www.lummustechnology.com/Process Technologies/Refining/Clean-Fuels/Gasoline-Alkylate-Production/AlkyClean%C2%AE

[19]   Salah, H.B.; Nancarrow, P.; Al-Othman, A. Ionic liquid-assisted refinery processes – A review and industrial perspective. *Fuel,* **2021,** *302,* 121195.
[http://dx.doi.org/10.1016/j.fuel.2021.121195]

[20]   Kore, R.; Scurto, A.M.; Shiflett, M.B. Review of Isobutane Alkylation Technology Using Ionic Liquid-Based Catalysts—Where Do We Stand? *Ind. Eng. Chem. Res.,* **2020,** *59*(36), 15811-15838.
[http://dx.doi.org/10.1021/acs.iecr.0c03418]

[21]   Wang, H.; Meng, X.; Zhao, G.; Zhang, S. Isobutane/butene alkylation catalyzed by ionic liquids: a more sustainable process for clean oil production. *Green Chem.,* **2017,** *19*(6), 1462-1489.
[http://dx.doi.org/10.1039/C6GC02791A]

[22]   Gan, P.; Tang, S. Research progress in ionic liquids catalyzed isobutane/butene alkylation. *Chin. J. Chem. Eng.,* **2016,** *24*(11), 1497-1504.
[http://dx.doi.org/10.1016/j.cjche.2016.03.005]

[23]   Díaz Velázquez, H.; Likhanova, N.; Aljammal, N.; Verpoort, F.; Martínez-Palou, R. New Insights into the Progress on the Isobutane/Butene Alkylation Reaction and Related Processes for High-Quality Fuel Production. A Critical Review. *Energy Fuels,* **2020,** *34*(12), 15525-15556.

[http://dx.doi.org/10.1021/acs.energyfuels.0c02962]

[24] Amarasekara, A.S. Acidic Ionic Liquids. *Chem. Rev.,* **2016**, *116*(10), 6133-6183.
      [http://dx.doi.org/10.1021/acs.chemrev.5b00763] [PMID: 27175515]

[25] Greaves, T.L.; Drummond, C.J. Protic ionic liquids: properties and applications. *Chem. Rev.,* **2008**, *108*(1), 206-237.
      [http://dx.doi.org/10.1021/cr068040u] [PMID: 18095716]

[26] Vafaeezadeh, M.; Alinezhad, H. Brønsted acidic ionic liquids: Green catalysts for essential organic reactions. *J. Mol. Liq.,* **2016**, *218*, 95-105.
      [http://dx.doi.org/10.1016/j.molliq.2016.02.017]

[27] Albright, L.F. Present and Future Alkylation Processes in Refineries. *Ind. Eng. Chem. Res.,* **2009**, *48*(3), 1409-1413.
      [http://dx.doi.org/10.1021/ie801495p]

[28] Wang, A.; Zhao, G.; Liu, F.; Ullah, L.; Zhang, S.; Zheng, A. Anionic Clusters Enhanced Catalytic Performance of Protic Acid Ionic Liquids for Isobutane Alkylation. *Ind. Eng. Chem. Res.,* **2016**, *55*(30), 8271-8280.
      [http://dx.doi.org/10.1021/acs.iecr.6b00768]

[29] Matuszek, K.; Chrobok, A.; Coleman, F.; Seddon, K.R.; Swadźba-Kwaśny, M. Tailoring ionic liquid catalysts: structure, acidity and catalytic activity of protonic ionic liquids based on anionic clusters, $[(HSO_4)(H_2SO_4)_x]^-$ (x = 0, 1, or 2). *Green Chem.,* **2014**, *16*(7), 3463-3471.
      [http://dx.doi.org/10.1039/C4GC00415A]

[30] Huang, Q.; Zhao, G.; Zhang, S.; Yang, F. Improved Catalytic Lifetime of H_2SO_4 for Isobutane Alkylation with Trace Amount of Ionic Liquids Buffer. *Ind. Eng. Chem. Res.,* **2015**, *54*(5), 1464-1469.
      [http://dx.doi.org/10.1021/ie504163h]

[31] Wang, L.; Zhao, G.; Yao, X.; Ren, B.; Zhang, S. Adamantane-Based Cation and $[MF_n]^-$ Anion Synergistically Enhanced Catalytic Performance of Sulfuric Acid for Isobutane Alkylation. *Ind. Eng. Chem. Res.,* **2017**, *56*(28), 7920-7929.
      [http://dx.doi.org/10.1021/acs.iecr.7b01192]

[32] Zhao, Y.; Li, T.; Meng, X.; Wang, H.; Zhang, Y.; Wang, H.; Zhang, S. Improvement of product distribution through enhanced mass transfer in isobutane/butene alkylation. *Chem. Eng. Res. Des.,* **2019**, *143*, 190-200.
      [http://dx.doi.org/10.1016/j.cherd.2018.12.014]

[33] Yu, F.L.; Gu, Y.L.; Gao, X.; Liu, Q.C.; Xie, C.X.; Yu, S.T. Alkylation of isobutane and isobutene catalyzed by trifluoromethanesulfonic acid-taurine deep eutectic solvents in polyethylene glycol. *Chem. Commun. (Camb.),* **2019**, *55*(33), 4833-4836.
      [http://dx.doi.org/10.1039/C9CC01254K] [PMID: 30950458]

[34] Estager, J.; Holbrey, J.D.; Swadźba-Kwaśny, M. Halometallate ionic liquids – revisited. *Chem. Soc. Rev.,* **2014**, *43*(3), 847-886.
      [http://dx.doi.org/10.1039/C3CS60310E] [PMID: 24189615]

[35] Kore, R.; Berton, P.; Kelley, S.P.; Aduri, P.; Katti, S.S.; Rogers, R.D. Group IIIA Halometallate Ionic Liquids: Speciation and Applications in Catalysis. *ACS Catal.,* **2017**, *7*(10), 7014-7028.
      [http://dx.doi.org/10.1021/acscatal.7b01793]

[36] Bui, T.L.T.; Korth, W.; Jess, A. Influence of acidity of modified chloroaluminate based ionic liquid catalysts on alkylation of iso-butene with butene-2. *Catal. Commun.,* **2012**, *25*, 118-124.
      [http://dx.doi.org/10.1016/j.catcom.2012.03.018]

[37] Pöhlmann, F.; Schilder, L.; Korth, W.; Jess, A. Liquid Phase Isobutane/2-Butene Alkylation Promoted by Hydrogen Chloride Using Lewis Acidic Ionic Liquids. *ChemPlusChem,* **2013**, *78*(6), 570-577.
      [http://dx.doi.org/10.1002/cplu.201300035]

[38] Xing, X.; Zhao, G.; Cui, J.; Zhang, S. Isobutane alkylation using acidic ionic liquid catalysts. *Catal.*

Commun., **2012**, *26*, 68-71.
[http://dx.doi.org/10.1016/j.catcom.2012.04.022]

[39] Cui, J.; de With, J.; Klusener, P.A.A.; Su, X.; Meng, X.; Zhang, R.; Liu, Z.; Xu, C.; Liu, H. Identification of acidic species in chloroaluminate ionic liquid catalysts. *J. Catal.,* **2014**, *320*, 26-32.
[http://dx.doi.org/10.1016/j.jcat.2014.09.004]

[40] Hu, P.; Wang, Y.; Meng, X.; Zhang, R.; Liu, H.; Xu, C.; Liu, Z. Isobutane alkylation with 2-butene catalyzed by amide-AlCl 3 -based ionic liquid analogues. *Fuel,* **2017**, *189*, 203-209.
[http://dx.doi.org/10.1016/j.fuel.2016.10.099]

[41] Wu, G.; Liu, Y.; Liu, G.; Hu, R.; Gao, G. Role of aromatics in isobutane alkylation of chloroaluminate ionic liquids: Insights from aromatic − ion interaction. *J. Catal.,* **2021**, *396*, 54-64.
[http://dx.doi.org/10.1016/j.jcat.2021.01.037]

[42] Timken, H.K.; Luo, H.; Chang, B-K.; Carter, E.; Cole, M. Chapter 2 ISOALKY™ Technology: Next-Generation Alkylate Gasoline Manufacturing Process Technology Using Ionic Liquid Catalyst. In: *Commercial Application of Ionic Liquids, Green Chemistry and Sustainable Technology*; Shiflett, M.B., Ed.; Springer, **2020**; pp. 33-47.
[http://dx.doi.org/10.1007/978-3-030-35245-5_2"10.1007/978-3-030-35245-5_2]

[43] Hu, P.; Wu, Z.; Wang, J.; Huang, Y.; Deng, Y.; Zhou, S. Analysis of long term catalytic performance for isobutane alkylation catalyzed by NMA–AlCl₃ based ionic liquid analog. *Chin. J. Chem. Eng.,* **2019**, *27*(8), 1857-1862.
[http://dx.doi.org/10.1016/j.cjche.2018.11.020]

[44] Hommeltoft, S.I.; Miller, S.J.; Pradhan, A. Process for Producing a Jet Fuel. U.S. Pat. 7,919,664, **2011**.

[45] Liu, Z.; Meng, X.; Zhang, R.; Xu, C.; Dong, H.; Hu, Y. Reaction performance of isobutane alkylation catalyzed by a composite ionic liquid at a short contact time. *AIChE J.,* **2014**, *60*(6), 2244-2253.
[http://dx.doi.org/10.1002/aic.14394]

[46] Zheng, W.; Li, D.; Sun, W.; Zhao, L. Multi-scale modeling of isobutane alkylation with 2-butene using composite ionic liquids as catalyst. *Chem. Eng. Sci.,* **2018**, *186*, 209-218.
[http://dx.doi.org/10.1016/j.ces.2018.04.043]

[47] Liu, Y.; Wu, G.; Pang, X.; Hu, R. Kinetics study on alkylation of isobutane with deuterated 2-butene in composite ionic liquids. *Chem. Eng. J.,* **2020**, *387*, 123407.
[http://dx.doi.org/10.1016/j.cej.2019.123407]

[48] Liu, Y.; Li, R.; Sun, H.; Hu, R. Effects of catalyst composition on the ionic liquid catalyzed isobutane/2-butene alkylation. *J. Mol. Catal. Chem.,* **2015**, *398*, 133-139.
[http://dx.doi.org/10.1016/j.molcata.2014.11.020]

[49] Harris, T.V.; Driver, M.; Elomari, S.; Timken, H-K.C. Alkylation Process Using an Alkyl Halide Promoted Ionic Liquid Catalyst. U.S. Pat. 20080142413A1. **2008**.

[50] Luo, H.; Ahmed, M. Ionic Liquid Catalyzed Alkylation Processes & Systems. U.S. Pat. 20130066130A1, **2013**.

[51] Elomari, S.; Timken, H-K.C. Isomerization of Butene in the Ionic Liquid-Catalyzed Alkylation of Light Isoparaffins and Olefins. U.S. Pat. 20080146858A1, **2008**.

[52] Ionikylation | Well Resources. https://www.wellresources.ca/ionikylation

[53] PetroChina's Ionikylation process based on ionic liquid--Lanzhou Greenchem ILs (Center for Greenchemistry and catalysis), LICP, CAS. Ionic Liquids. http://www.ionike.com/en/application/2014-04-24/40.html

[54] Safe and sustainable alkylation: Performance and update on composite ionic liquid alkylation technology. https://www.hydrocarbonprocessing.com/magazine/2020/april-2020/special-focus-cl-an-fuels/safe-and-sustainable-alkylation-performance-and-update-on-composite-i-nic-liquid-alkylation-technology

[55] *Sinopec successfully starts-up largest composite ionic liquid alkylation unit,* https://www.hydrocarbonprocessing.com/news/2019/04/sinopec-successfully-starts-up-largest-composite-ionic-liquid-alkylation-unit

[56] Bui, T.L.T.; Korth, W.; Aschauer, S.; Jess, A. Alkylation of isobutane with 2-butene using ionic liquids as catalyst. *Green Chem.,* **2009**, *11*(12), 1961-1967.
[http://dx.doi.org/10.1039/b913872b]

[57] Liu, S.; Wang, Z.; Li, K.; Li, L.; Yu, S.; Liu, F.; Song, Z. Brønsted-Lewis acidic ionic liquid for the "one-pot" synthesis of biodiesel from waste oil. *J. Renew. Sustain. Energy,* **2013**, *5*(2), 023111.
[http://dx.doi.org/10.1063/1.4794959]

[58] Liu, S.; Chen, C.; Yu, F.; Li, L.; Liu, Z.; Yu, S.; Xie, C.; Liu, F. Alkylation of isobutane/isobutene using Brønsted–Lewis acidic ionic liquids as catalysts. *Fuel,* **2015**, *159*, 803-809.
[http://dx.doi.org/10.1016/j.fuel.2015.07.053]

[59] Kumar, P.; Vermeiren, W.; Dath, J-P.; Hoelderich, W.F. Production of Alkylated Gasoline Using Ionic Liquids and Immobilized Ionic Liquids. *Appl. Catal. Gen,* **2006**, *304*, 131-141.
[http://dx.doi.org/10.1016/j.apcata.2006.02.030]

[60] Jin, K.; Zhang, T.; Yuan, S.; Tang, S. Regulation of isobutane/1-butene adsorption behaviors on the acidic ionic liquids-functionalized MCM-22 zeolite. *Chin. J. Chem. Eng.,* **2018**, *26*(1), 127-136.
[http://dx.doi.org/10.1016/j.cjche.2017.05.023]

[61] Liu, Y.; Liu, G.; Wu, G.; Hu, R. Alkylation of Isobutane and 2-Butene in Rotating Packed-Bed Reactors: Using Ionic Liquid and Solid Acid as Catalysts. *Ind. Eng. Chem. Res.,* **2020**, *59*(33), 14767-14775.
[http://dx.doi.org/10.1021/acs.iecr.0c02520]

CHAPTER 10

Other Applications of ILs in the Petroleum Industry

Abstract: In this chapter, we will discuss some other applications that are of great importance in the oil industry and in which ILs have played an important role; however, the IMP has not yet ventured into these topics, at least not with the use of ILs. The topics discussed in this chapter are the separation of light hydrocarbons, separation and extraction of aromatic and aliphatic hydrocarbons, extraction of bitumen from oil sand, and application of ILs in shale stabilization processes.

Keywords: Aliphatic hydrocarbons, Aromatic hydrocarbons, Bitumen extraction, Extraction, Ionic liquids, Separation.

INTRODUCTION

The separation of light hydrocarbons is a subject of great importance in the oil industry. This process refers to the room-temperature separation of gaseous or liquid hydrocarbons, saturated and unsaturated, including olefins, dienes, and alkynes. The separation of these products is of interest to add value to these pure hydrocarbons or to transform them into other added-value compounds; for example, the synthesis of polyethylene and propylene requires the use of high purity monomers. Also, the separation of mixtures of hydrocarbons is important both because of their use as end products or for the preparation of feedstocks employed in clean energy processes [1].

One of the alternatives for light hydrocarbon separation is by means of cryogenic energy [2]. Extractive distillation [3, 4] and separation with the aid of porous materials [5] and metal-organic frameworks (MOFs) have also been studied [6].

Separation of Light Hydrocarbons Employing ILs

A novel option that has been intensively studied for the separation of light hydrocarbons in recent years is the use of ILs as effective solvents, as they represent environmentally friendly technology [7].

Some results of the successful application of ILs for the selective separation of different light hydrocarbons are described below in chronological order. In 2008, Ortiz *et al.* studied the selective adsorption of propylene from its mixtures with propane by chemical complexation with silver ions in IL solutions of 1-butyl-3-methylimidazolium tetrafluoroborate (silver salt concentration = 0.25 M in form of $AgBF_4$) as functions of temperature and pressure. In all cases, gas solubilities increased with pressure and decreased with the system temperature.

The process with silver in the presence of IL proved much more effective for the selective adsorption of propylene in the presence of propane due to the fact that the silver cation becomes chemically more active forming silver-olefin complexes. The complete regeneration of the reaction medium containing the IL can be carried out at room temperature, 800 rpm of stirring rate and at 20-mbar vacuum for 3 h.

However, despite the silver-containing ILs provide tailorable selectivity to olefins, in our consideration, these results would be difficult to produce on a large scale due to the high cost of the $AgBF_4$ reagent involved in the process [8].

In 2010, researchers at the German University of Erlangen-Nuremberg carried out the screening of a family of ILs with imidazolium cation and one ammonium-type IL in order to find the most suitable prototype for propene/propane separation using the extractive distillation method (Fig. **10.1**). This study showed that the IL structure significantly influences the selective propene adsorption. In general, all the tested ILs preferred the selective propene separation in the presence of propane and it was observed that the gas solubility increased with the growing length of the alkyl chain linked to the imidazolium ring and in the same sense, the ability of the IL to separate the gases decreased. The best performing prototype was the IL containing the $[B(CN)_4]$ anion, which is highlighted in red in Fig. (**10.1**) [9].

In 2011, Hu *et al.* published a review featuring a critical discussion of the current state of knowledge of the key factors influencing the solubility of gases in ILs, including sample purity, experimental methodology, "molecular" characteristics of ILs, temperature and pressure. In this review, it was concluded that the presence of ion conformational equilibria and the formation of ionic and non-polar domains in the ILs influence significantly the solubility of gases in ILs with physical adsorption. The work is mainly focused on gases such as CO_2, SO_2, N_2, O_2 and CH_3 and on the separation of their mixtures.

The increased flexibility of the anion structures, and the substituents on the head and tail of the cation contribute to increase the free volume and decrease the cavity formation energy. The joint use of these findings is of great importance when we note the fact that the use of ILs will depend on whether their viscosity and chemical stability are suitable for a given application [10].

Cation   Anion

R—N⟨+⟩N—R₁

R = Me, R₁ = Bu	Tf₂N
R = Me, R₁ = Pr	Tf₂N
R = Me, R₁ = Et	Tf₂N
R = Me, R₁ = Me	Tf₂N
R = Bu, R₁ = Bu	Tf₂N
R = Pr, R₁ = Bu	Tf₂N
R = Et, R₁ = Bu	Tf₂N
R = Et, R₁ = Pr	Tf₂N
R = Me(aryl), R₁ = Me	Tf₂N
R = Me, R₁ = Bu	DCA
R = Me, R₁ = Et	DCA
R = Oc, R₁ = Me	Tf₂N
R = Oc, R₁ = Me	C(CN)₃
R = C₆CN, R₁ = Oc	Tf₂N
R = C₆CN, R₁ = Me	Tf₂N
R = Me, R₁ = Et	C(CN)₃
R = C₆CN, R₁ = Me	C(CN)₃
R = Et, R₁ = Me	B(CN)₄

Tf_2N^- C_8H_{17}
$H_3C{-}\overset{+}{N}{-}C_8H_{17}$
                C_8H_{17}

Fig. (10.1). ILs evaluated in the propene/propane separation using the extractive distillation method (Mokrushin *et al.*, 2010).

In 2013, Xing *et al.* [11] studied the solubility of ethylene and ethane in three ILs at 303.15 K: 1-butyl-3-methylimidazolium bis(trifluoromethylsulfonyl)imide ([BMIM][NTf₂]), 1-butyronitrile bis((trifluoromethyl)sulfonyl)imide ([CPMIM][NTf₂]), and dual-functionalized [(CP)₂IM][NTf₂] (Fig. **10.2**). It was found that the symmetric IL showed the highest separation selectivity to ethylene/ethane on [(CP)₂IM][NTf₂]. Theoretical calculations employing

COSMO-RS were carried out to explain the experimental results and it was shown that the polarity of the IL and its mismatch interaction with the gases were the main factors determining the solubilities of ethylene and ethane in it.

Also, in 2013, Liu *et al.* described in two papers the high solubility of light hydrocarbons in the ILs tetrabutylphosphonium bis(2,4,4-trimethylpentyl) phosphinate, and 1-ethyl-3- methylimidazolium bis(trifluoromethylsulfonyl) imide, which showed high solubility for methane, ethane, ethylene and propane, and trihexyl tetradecylphosphonium bis(2,4,4-trimethylpentyl) phosphinate displayed solubilities of paraffins (ethane and propane) that were higher than those of the corresponding olefins (ethylene and propylene) [12, 13].

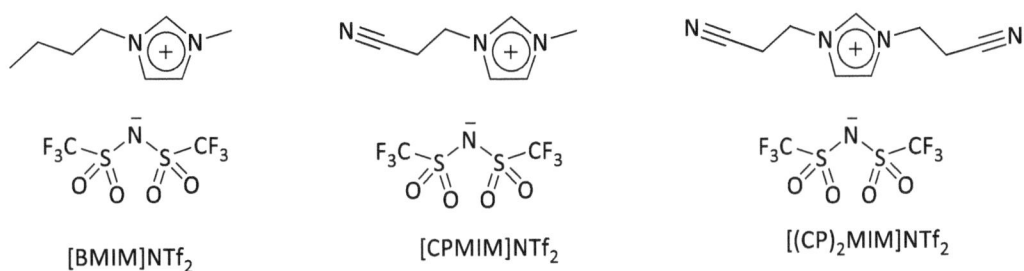

Fig. (10.2). Structure of the ILs studied in this study [11].

Moura *et al.*, in 2015, also carried out the experimental screening of IL absorption of ethane and ethylene as a function of temperature. The solubility of ethylene was higher than that of ethane in the studied ILs and varied from 18.41×10^{-3} in $[C_1C_8IM][NTf_2]$ at 303.17 K to 1.603×10^{-3} in $[C_1C_4IM][DCA]$ at 343.61 K. The addition of unsaturated groups such as propyn-3-yl and cyanopropyl to the alkyl chain of the cation in the imidazolium ILs caused a decrease in the solubility of ethane and ethylene compared to the IL $[BMIM]NTf_2$, which was taken as the reference compound [14].

In 2016, the same research group reviewed the state of the art of the selective gas adsorption by ILs and showed that the adsorption capacity of an imidazolium-based IL was doubled with the addition of a copper(II) salt [15]. They also found that the solubility trends of ethane, ethylene, propane and propylene in ILs increased with the size of the non-polar domains of the cation or anion of the IL, suggesting that the solubility of these gases in the group of studied ILs is ruled by nonspecific interactions. (Fig. **10.3**) According to this review, the authors defined that solubility in terms of cations followed the order:

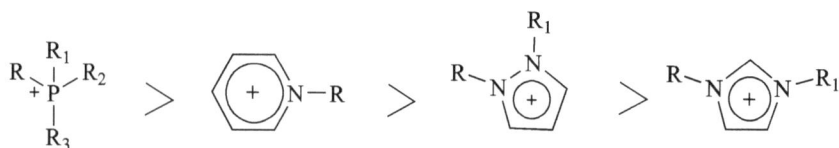

Fig. (10.3). Solubility order of ethane, ethylene, propane and propylene in terms of cations according to Moura *et al.* (2016) [15].

The solubility of ILs such as imidazolium-based lipid ILs using ethane and ethylene [16], and a theoretical (by COSMO-RS) and experimental screening study on the separation of the same light hydrocarbons were also described in 2015. Based on the screening of 420 ILs by COSMO-RS calculations, tetraalkylphosphonium-based ILs with long-chain carboxylate anions were designed for C_2H_2/C_2H_4 separation and experimentally the IL tetrabutylphosphonium pentacaboxylate ($[P_{4444}][C_5COO]$) showed excellent absorption performance for C_2H_2 with separation selectivity up to 21.4 at 298.1 K [17].

Recently, Zhang *et al.* found that also ILs with asymmetric phosphonium cation and long- chain-carboxylate-type anions showed very good performance in the separation of light gases by displaying high solubility and low viscosity for light hydrocarbon separations. It was elicited that the asymmetric anion containing a short-chain carboxylate anion reduced significantly the IL viscosity whereas the increase in the anion chain length produced better solubility. Thus, the IL tributylethylphosphonium octadecylcarboxylate displayed high solubility, high thermal stability and excellent selectivity: propane/methane of 13.4, propane/nitrogen of 38.9, ethane/methane of 5.6 and ethane/nitrogen of 16.2 [18].

More recently, [BMIM]OAc supported on a hydrothermally stable MOFs, namely MIL-101(Cr), was utilized to separate efficiently acetylene from ethylene. The composite IL-MOF increased remarkably the adsorption selectivity of acetylene/ethylene from 3.0 to 30 in comparison to MIL-101(Cr) [19].

Separation of Aromatic and Aliphatic Hydrocarbons Using ILs

In addition to the separation of light hydrocarbons and that of contaminating hydrocarbon compounds such as nitrogenous and sulfur compounds, which was discussed in Chapter 2 of this compendium, one of the most important separation processes in the Oil Industry is the separation of aliphatic and aromatic hydrocarbons.

Depending on the origin of the feedstock, the total aromatic content of the feed streams can vary from 25 to 75%. The separation of aromatic hydrocarbons is of great importance, since these compounds play a very important role in the petrochemical industry and aromatic compounds can inhibit the desulfurization and denitrogenation processes; in addition, the aromatic compounds themselves cause environmental problems when burned [20].

The separation of aromatic hydrocarbons (benzene, toluene, ethyl benzene and xylenes) from C_4 to C_{10} aliphatic hydrocarbon mixtures is challenging since these hydrocarbons have boiling points within a close range and several combinations form azeotropes. The conventional processes for the separation of these aromatic and aliphatic hydrocarbon mixtures are liquid extraction, suitable for the interval ranging from 20 to 65 wt.% of aromatic content, extractive distillation for the 65–90 wt.% interval of aromatics and azeotropic distillation for high aromatic content, > 90 wt.%.

Previously employed methods in the extraction with conventional solvents such as propylene carbonate [21], tetraethyleneglycol [22], and sulfolane [23] were investigated for the separation of aliphatic/aromatic compounds from naphtha, but these processes were not economically feasible since additional separation steps are required to purify the raffinate.

The first work describing the successful use of ILs in the extraction of low concentrations of aromatic compounds (less than 20%) was described by Meindersma and Haan in 2005. In this work, they tested several ILs, being [BMM]BF_4, [MeBuPy]BF_4, and [MeBuPy]CH_3SO_4 at 40 °C the ones that showed the best performance in the extraction of toluene/heptane and in all cases, these compounds showed better selectivity (S) and distribution coefficient (D) than sulfolane. The IL 3-methyl-N-butylpyridinium tetrafluoroborate ([MeBuPy]BF_4) displayed the best combination of both a high toluene distribution coefficient and a high toluene/heptane selectivity [24].

Then, in 2006, the same authors studied another group of ILs at 40 and 75 °C, and newly [MeBuPy]BF_4 was the most suitable compound for the extraction of toluene/heptane as model mixture. The S to the toluene/heptane separation using ([MeBuPy]BF_4) increased by decreasing the toluene content in the feed: from 12 to 53 (from 95 to 5% of toluene in the feed) at 40°C and from 17 to 39 (from 65 to 10% of toluene in the feed) at 75°C. [MeBuPy]BF_4 (Fig. **10.4**) was selected for a study in a pilot plat [25].

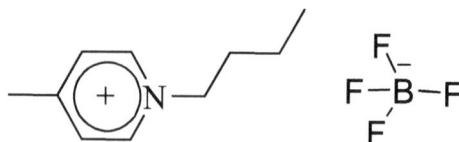

Fig. (10.4). Structure of the IL 4-methyl-N-butylpyridinium tetrafluoroborate [MeBuPy]BF$_4$.

In 2008, also Meindersma and Haan [26] developed a conceptual process design for aromatic/aliphatic separation with the IL [MeBuPy]BF$_4$, which was developed using the flow sheeting program in ASPEN Plus 12.1 that resulted in a process with a positive profit margin, which was also demonstrated with an actual pilot plant extraction column. The authors also found that a high aromatic distribution coefficient is the key factor to a feasible extraction process with ILs.

The use of [MeBuPy]BF$_4$ as a viable alternative for the separation of low aromatic contents in naphtha was further confirmed by means of a computer-assisted methodology that integrates molecular modeling and process simulation *via* COSMO-based thermodynamic models in Aspen Plus by Riva *et al.*, in 2016 [27].

Zhang *et al.* evaluated a series of chloroaluminate ILs, being the Lewis acid 1-butyl-e-methylimidazolium chloride aluminate ([BMIM]Cl/AlCl$_3$ 2.0 the IL with the best extractive performance. These ILs can be regenerated through vacuum distillation [28].

In later years, other ILs such as 1-butyl-1-methylpyrrolidinium bis (trifluoromethylsulfonyl) imide [BuMePyr][NTf$_2$] [29], a mixture of 4-ethyl-*N*-methylpyridinium bis(trifluoromethylsulfonyl)imide and 1-ethyl, 3-methyl dicyanamide {[4EtMePy][Tf$_2$N] (0.3) + [EMIM][DCA] (0.7)} [30], 1-methyl-3-octylimidazolium tetrafluoroborate [31], and theoretically (Conductor-like Screening Model for Real Solvents (COSMO-RS) module) and experimentally validated by liquid–liquid extraction compounds like 1-ethyl-3-methyli-midazolium acetate ([C$_2$MIM][Ac]), 1-ethyl-3-methylimidazolium dicyanamide ([C$_2$MIM][N(CN)$_2$]), 1-ethyl-3-methylimidazolium thiocyanate ([C$_2$MIM][SCN]) and 1-ethyl-3-methylimidazolium bis(trifluoromethylsulfonyl) imide ([C$_2$MIM] [Tf$_2$N]) [32] have shown good performance in the liquid-liquid extraction of the aromatic/aliphatic mixture.

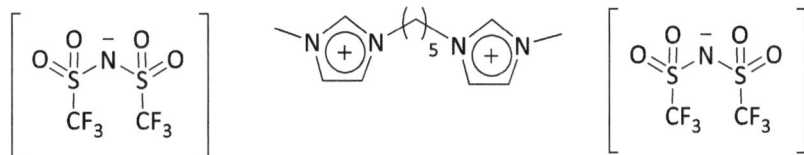

Fig. (10.5). Dicationic ILs used in the extraction of aromatics from aliphatics.

More recently, Yao *et al.* (2019) [33] studied the extraction of aromatics/aliphatics using a hydrophobic dicationic IL adjusted with low-content water. According to these authors, the previous studies with ILs could reach neither high purity nor extraction efficiency of aromatic products. In their study, the authors found that the dicationic ILs ($[C_5(MIM)_2][NTf_2]_2$, Fig. **10.5**) could dissolve a small amount of water with solubility of 2.24 wt.% at 303.15 K, avoiding the negative effect on the mass transfer of the extraction process due to the high viscosity of the dicationic IL. With these ILs both D and S were high for separation of aliphatic/aromatic mixtures when a small amount of water was dissolved in the DIL. The highest selectivity and distribution coefficient for toluene were 45.7, 0.86 (water content: 1.10 wt.%) and 41.5, 0.75 (water content: 1.59 wt.%), respectively. By increasing the carbon chain of aliphatics, both D_{tol} and D_{ali} decreased, but S increased in the presence of water.

ILs in Shale Stabilization Processes

Some of the oil is trapped in the pores of low-permeability rock formations called oil shales located underground. They are usually found at depths between 1,000 and 5,000 meters.

Due to the low permeability of shales, the extraction of hydrocarbons requires the use of hydraulic fracturing or fracking. This technique starts with the drilling of a vertical well until the formation containing gas or oil is reached. A series of horizontal boreholes are then drilled into the shale, which can extend for several kilometers in various directions. Through these horizontal wells, the rock is fractured by injecting a mixture of water, sand and chemicals at high pressure, which forces the hydrocarbons to flow out of the pores. But this flow diminishes very soon, so it is necessary to drill new wells to maintain the production of the reservoirs [34].

Hydraulic fracturing, in addition to high water consumption, it creates swelling, dispersion, and solids buildup in the mud. In some cases, hydration can lead to hole collapse, tight holes, stuck, and pore plugging. Many studies have been carried out to resolve these undesirable problems in shale gas production. For years, the most common solution has been the use of additives, primarily sodium and potassium chloride, surfactants, biosurfactants and polymers that retard the hydration of clays, however, the use of these compounds has its limitations and ecological problems [35].

ILs have shown to inhibit effectively shale swelling by being adsorbed successfully on clay minerals, creating a hydrophobic protection around the clay rocks [36].

In 2015, an environmentally friendly and biodegradable shale inhibitor based on chitosan quaternary ammonium salt IL (HTCC) was evaluated by An *et al.* In this work, the inhibition was evaluated by the linear swelling test, mud making test and rolling recovery. The results indicated that the inhibition of HTCC was better than that displayed by polyether amino, which can be used widely in the oil flied as an excellent shale inhibitor [37].

Yang, *et al.* studied the application of the IL monomeric 1-vinyl-3-ethylimidazolium bromide ([VeEIM]Br) and of the corresponding homopolymer poly(vinylimidazole) bromide (PVBr) as shale hydration inhibitors.

PVBr was synthesized by radical polymerization of [VeEIM]Br using 2,2′-azobis(2-methylpropionamidine) dihydrochloride (V-50) as initiator (Fig. **10.6**).

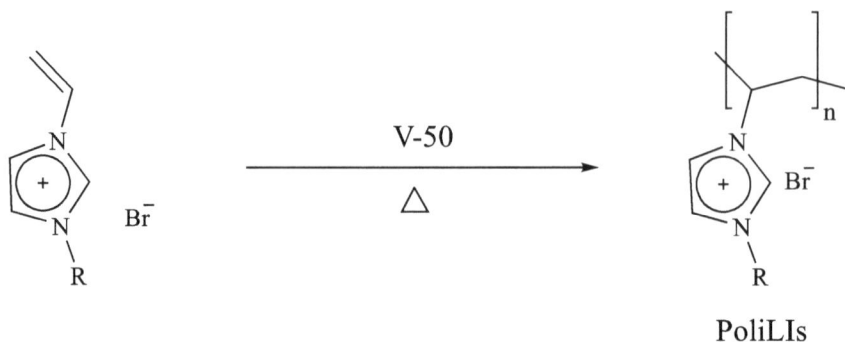

Fig. (10.6). Synthesis of PVBr from [VeEIM]Br.

[VeEIM]Br and PVBr displayed better thermal stability, beneficial to high-temperature drilling, than KCl and 2,3-epoxypropyltrimethylammonium chloride (EPTAC), which are known to be high performance additives. By the polymerization method, PVBr can encapsulate bentonite by multi adsorption sites, and can be used as a sealing additive to prevent water flowing into shales [38].

Also, Yang, in 2019, studied the effect of the alkyl chain length on the shale hydration inhibitive performance of a vinylimidazolium-based IL. A series of vinylimidazolium-based ILs with different alkyl saturated chain lengths, from C2 to C14, was prepared and evaluated as inhibitors in drilling fluids to inhibit the hydration, swelling, and dispersion of Na-BT. All the ILs displayed good thermal stability, beneficial to high-temperature drilling. For inhibition performance, ILs with the ethyl group showed superior ability inhibiting hydration, minimizing the degree of linear swelling, abating the dispersion of sodium bentonite (Na-BT), and Na-BT suspension at temperatures of 25, 60, 80 and 120 °C [39] (Fig. **10.7**).

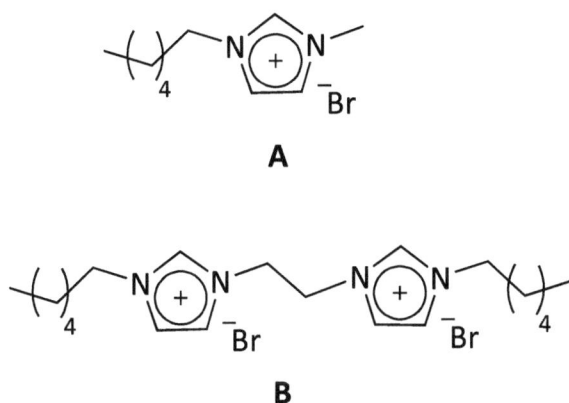

A

B

Fig. (10.7). ILs evaluated as shale inhibitors by Jia *et al*. [41].

Other ILs that have shown good performance as environmentally friendly shale inhibitors for water-based drilling fluids are quaternary ammonium salts (2,3-epoxypropyltrimethylammonium chloride (EPTAC)-modified gelatin) [40], and 1-hexyl-3-methylimidazolium bromide (**A**) and 1,2-bis(3- hexylimidazolium-1-yl) ethane bromide (**B**) [41].

IL-assisted Bitumen Extraction from Oil Sand

Bitumen is a complex mixture, containing a high proportion of poorly soluble asphaltenes. Given the large reserves of fossil fuel in the form of bitumen, there is great interest in extracting oil sand from bitumen in an environmentally friendly, sustainable and potentially economical way. To obtain one barrel of oil, about 2 tons of oil sand are required. The extraction of oil sand from bitumen and clay with hot water has been the most used method; however, this method represents a great environmental problem since it requires a very large volume of water and high energetic expense. In this sense, ILs have shown great potential to meet this technological challenge as shown in the following examples [42].

In 2010, Painter, Williams and Mannebach studied three ILs for recovering bitumen from oil or tar sands [43]. Separations of bitumen were performed by placing a sample of tar sands, toluene, and an IL in the proportions 1:2:3 per weight either in glass vials or centrifuge tubes. The ILs 1-butyl-3-metyl-imidazolium trifluoromethanesulfonate, 1-butyl-2,3-dimethyl-imidazolium tetrafluoroborate, and 1-butyl-2,3-dimethylimidazolium trifluoromethanesulfonate studied are shown in Fig. (**10.8**).

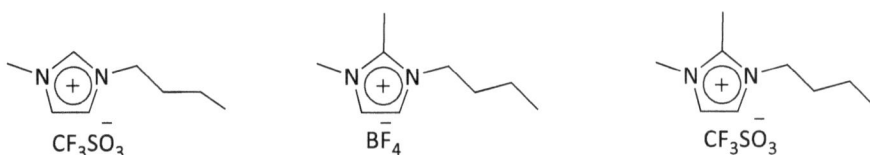

Fig. (10.8).   Structure of ILs studied for bitumen extraction from oil sand by Painter, Williams and Mannebach [43].

According to the preliminary experiments carried out in this study, the ILs had yields above 90% in bitumen extraction with the advantage that at this first stage, the use of water was not required. Water was subsequently used to remove the IL from the residual sand and clays and it was easily separated by distillation from the IL since ILs have practically zero vapor pressure. ILs with methyl substituent at carbon 2 of imidazole showed much better performance. No IL contamination was detected in the residual sand and clays and the bitumen produced through this process was free of both clay fines and residual IL.

Li *et al.* studied two ILs: 1-butyl-2,3-dimethylimidazolium tetrafluoroborate ([BMMIM]BF_4 and ethyl-3-methylimidazolium tetrafluoroborate ([EMIM]BF_4). The low viscosity [EMIM]BF_4 displayed the best performance improving the recovery of bitumen from oil sands by solvent extraction, using a solvent composed of *n*-heptane and acetone. The results showed that [EMIM]BF_4 increased the bitumen recovery up to 95% at room temperature. The presence of clay fines in the recovered bitumen was much lower than in the case of solvent extraction without IL. No organic residues were observed in the spent sands. This technology avoided the glue water problem because only a small amount of it was used to recycle the IL, and the organic solvent could be easily regenerated by distillation.

Acetone played an important role in the extractive power of the mixture by significantly decreasing the viscosity of [EMIM]BF_4 and the amount of IL required in the process. The optimal conditions for this procedure were acetone/n-heptane ratio (2:6, v/v), mixing time (10 min), stirring speed (450 rpm), temperature (25°C), and organic solvents to oil sands ratio of 4:1, v/w, which were determined for implementing this technology [44].

Likewise, the same IL [EMIM]BF_4 was recently investigated for its effectiveness enhancing the oil recovery from hazardous crude oil tank bottom sludge using a solvent [45].

In 2017, Tourvieille *et al.* studied several ILs for bitumen extraction from oil sand. The ILs that met the eligibility criteria for this application were selected: very low viscosity (μ < 50 mPa.s), high hydrolysis stability and good miscibility with water. Based on these criteria, the following ILs were selected: 1-Allyl-3-methylimidazolium dicyanamide: [AMIM][DCA], ethylammonium nitrate: [EtNH$_3$][NO$_3$], 1-ethyl-3-methylimidazolium dicyanamide: [EMIM][DCA], and 1-ethyl-3-methylimidazolium thiocyanate: [EMIM][SCN], which were compared to a previously reported IL with good performance for this application: 1-butyl-2, 3-dimethylimidazolium tetrafluoroborate: [BMMIM][BF$_4$]. An innovative continuous flow microcell coupled to near infrared (NIR) spectroscopy was developed to monitor online the IL-assisted bitumen extraction kinetics along with a quantitative reconstruction strategy using the projection to latent structure (PLS) method. The choice of solvent, rather than IL, was the main factor influencing the bitumen recovery while the NIR extraction kinetic tests revealed that the IL viscosity drastically reduced the bitumen recovery rate constant. The best results were obtained with [EtNH$_3$][NO$_3$] [46].

Extraction of bitumen has also been investigated with deep eutectic solvent (DES) IL analogue. An environmentally friendly DES obtained by mixtures of choline chloride and urea (ChCl/U) mixed with naphtha as diluent to reduce bitumen viscosity was used for the extraction of bitumen for sand oil. Extraction yields above 80% of bitumen were obtained for the initial 10 min of extraction using 75% of ChCl/U solution. A simulated countercurrent extraction of an Alberta oil sand sample gave in-excess extraction yields of 90% [47].

Other applications in which ILs have shown good performance are flow assurance, avoiding wax [48], an asphaltene deposition [49 - 51].

CONCLUDING REMARKS

As we have seen in the present chapters, ILs are being investigated in another series of applications of great impact in the oil industry. Although research in these areas is still at the laboratory level in most cases, it is expected that in the near future large-scale applications in processes such as the separation and extraction of aromatic and aliphatic hydrocarbons, bitumen extraction and the extraction of hydrocarbons will be considered. Although the use of fluorinated ILs in separation processes was booming at the beginning of the 20th century, however, currently the interest is in ILs with organic acid-based anions, mostly for separation of aromatic and aliphatic hydrocarbons.

REFERENCES

[1]   Barnett, B.R.; Gonzalez, M.I.; Long, J.R. Recent Progress Towards Light Hydrocarbon Separations Using Metal–Organic Frameworks. *Trends Chem.,* **2019**, *1*(2), 159-171.
        [http://dx.doi.org/10.1016/j.trechm.2019.02.012]

[2]   Gao, T.; Lin, W.; Gu, A. Improved processes of light hydrocarbon separation from LNG with its cryogenic energy utilized. *Energy Convers. Manage.,* **2011**, *52*(6), 2401-2404.
        [http://dx.doi.org/10.1016/j.enconman.2010.12.040]

[3]   Lei, Z.; Li, C.; Chen, B. Extractive Distillation: A Review. *Separ. Purif. Rev.,* **2003**, *32*(2), 121-213.
        [http://dx.doi.org/10.1081/SPM-120026627]

[4]   Yang, X.; Yin, X.; Ouyang, P. Simulation of 1,3-Butadiene Production Process by Dimethylfomamide Extractive Distillation. *Chin. J. Chem. Eng.,* **2009**, *17*(1), 27-35.
        [http://dx.doi.org/10.1016/S1004-9541(09)60028-8]

[5]   Cui, X.; Chen, K.; Xing, H.; Yang, Q.; Krishna, R.; Bao, Z.; Wu, H.; Zhou, W.; Dong, X.; Han, Y.; Li, B.; Ren, Q.; Zaworotko, M.J.; Chen, B. Pore chemistry and size control in hybrid porous materials for acetylene capture from ethylene. *Science,* **2016**, *353*(6295), 141-144.
        [http://dx.doi.org/10.1126/science.aaf2458] [PMID: 27198674]

[6]   Bao, Z.; Chang, G.; Xing, H.; Krishna, R.; Ren, Q.; Chen, B. Potential of microporous metal–organic frameworks for separation of hydrocarbon mixtures. *Energy Environ. Sci.,* **2016**, *9*(12), 3612-3641.
        [http://dx.doi.org/10.1039/C6EE01886F]

[7]   Huang, Y.; Zhang, Y.; Xing, H. Separation of light hydrocarbons with ionic liquids: A review. *Chin. J. Chem. Eng.,* **2019**, *27*(6), 1374-1382.
        [http://dx.doi.org/10.1016/j.cjche.2019.01.012]

[8]   Ortiz, A.; Ruiz, A.; Gorri, D.; Ortiz, I. Room temperature ionic liquid with silver salt as efficient reaction media for propylene/propane separation: Absorption equilibrium. *Separ. Purif. Tech.,* **2008**, *63*(2), 311-318.
        [http://dx.doi.org/10.1016/j.seppur.2008.05.011]

[9]   Mokrushin, V.; Assenbaum, D.; Paape, N.; Gerhard, D.; Mokrushina, L.; Wasserscheid, P.; Arlt, W.; Kistenmacher, H.; Neuendorf, S.; Göke, V. Ionic Liquids for Propene-Propane Separation. *Chem. Eng. Technol.,* **2010**, *33*, 63-73.
        [http://dx.doi.org/10.1002/ceat.200900343]

[10]   Hu, Y.F.; Liu, Z.C.; Xu, C.M.; Zhang, X.M. The molecular characteristics dominating the solubility of gases in ionic liquids. *Chem. Soc. Rev.,* **2011**, *40*(7), 3802-3823.
        [http://dx.doi.org/10.1039/c0cs00006j] [PMID: 21412560]

[11]   Xing, H.; Zhao, X.; Li, R.; Yang, Q.; Su, B.; Bao, Z.; Yang, Y.; Ren, Q. Improved Efficiency of Ethylene/Ethane Separation Using a Symmetrical Dual Nitrile-Functionalized Ionic Liquid. *ACS Sustain. Chem.& Eng.,* **2013**, *1*(11), 1357-1363.
        [http://dx.doi.org/10.1021/sc400208b]

[12]   Liu, X.; Afzal, W.; Prausnitz, J.M. Solubilities of Small Hydrocarbons in Tetrabutylphosphonium Bis(2,4,4-trimethylpentyl) Phosphinate and in 1-Ethyl-3-methylimidazolium

Bis(trifluoromethylsulfonyl)imide. *Ind. Eng. Chem. Res.,* **2013**, *52*(42), 14975-14978.
[http://dx.doi.org/10.1021/ie402196m]

[13]   Liu, X.; Afzal, W.; Yu, G.; He, M.; Prausnitz, J.M. High solubilities of small hydrocarbons in trihexyl
        tetradecylphosphonium bis(2,4,4-trimethylpentyl) phosphinate. *J. Phys. Chem. B,* **2013**, *117*(36),
        10534-10539.
        [http://dx.doi.org/10.1021/jp403460a] [PMID: 23947453]

[14]   Moura, L.; Darwich, W.; Santini, C.C.; Costa Gomes, M.F. Imidazolium-based ionic liquids with
        cyano groups for the selective absorption of ethane and ethylene. *Chem. Eng. J.,* **2015**, *280*, 755-762.
        [http://dx.doi.org/10.1016/j.cej.2015.06.034]

[15]   Moura, L.; Santini, C.C.; Costa Gomes, M.F. Gaseous Hydrocarbon Separations Using Functionalized
        Ionic Liquids. *Oil Gas Sci. Technol. –. Rev. IFP Energ. Nouv.,* **2016**, *71*, 23.
        [http://dx.doi.org/10.2516/ogst/2015041]

[16]   Green, B.D.; O'Brien, R.A.; Davis, J.H., Jr; West, K.N. Ethane and Ethylene Solubility in an
        Imidazolium-Based Lipidic Ionic Liquid. *Ind. Eng. Chem. Res.,* **2015**, *54*(18), 5165-5171.
        [http://dx.doi.org/10.1021/ie505071t]

[17]   Zhao, X.; Yang, Q.; Xu, D.; Bao, Z.; Zhang, Y.; Su, B.; Ren, Q.; Xing, H. Design and screening of
        ionic liquids for C_2H_2/C_2H_4 separation by COSMO-RS and experiments. *AIChE J.,* **2015**, *61*(6),
        2016-2027.
        [http://dx.doi.org/10.1002/aic.14782]

[18]   Zhang, Y.; Zhao, X.; Yang, Q.; Zhang, Z.; Ren, Q.; Xing, H. Long-Chain Carboxylate Ionic Liquids
        Combining High Solubility and Low Viscosity for Light Hydrocarbon Separations. *Ind. Eng. Chem.
        Res.,* **2017**, *56*(25), 7336-7344.
        [http://dx.doi.org/10.1021/acs.iecr.7b00660]

[19]   Wang, J.; Xie, D.; Zhang, Z.; Yang, Q.; Xing, H.; Yang, Y.; Ren, Q.; Bao, Z. Efficient adsorption
        separation of acetylene and ethylene *via* supported ionic liquid on metal-organic framework. *AIChE J.,*
        2017, *63*(6), 2165-2175.
        [http://dx.doi.org/10.1002/aic.15561]

[20]   Sharma, M.; Sharma, P.; Kim, J.N. Solvent extraction of aromatic components from petroleum derived
        fuels: a perspective review. *RSC Advances,* **2013**, *3*(26), 10103-10126.
        [http://dx.doi.org/10.1039/c3ra00145h]

[21]   Ali, S.H.; Lababidi, H.M.S.; Merchant, S.Q.; Fahim, M.A. Extraction of aromatics from naphtha
        reformate using propylene carbonate. *Fluid Phase Equilib.,* **2003**, *214*(1), 25-38.
        [http://dx.doi.org/10.1016/S0378-3812(03)00323-6]

[22]   Wang, W.; Gou, Z.; Zhu, S. Liquid–Liquid Equilibria for Aromatics Extraction Systems with
        Tetraethylene Glycol. *J. Chem. Eng. Data,* **1998**, *43*(1), 81-83.
        [http://dx.doi.org/10.1021/je970152i]

[23]   Choi, Y.J.; Kwon, T.I.; Yeo, Y.K. Optimization of the sulfolane extraction plant based on modeling
        and simulation. *Korean J. Chem. Eng.,* **2000**, *17*(6), 712-718.
        [http://dx.doi.org/10.1007/BF02699122]

[24]   Wytze Meindersma, G.; Podt, A.J.G.; de Haan, A.B. Selection of ionic liquids for the extraction of
        aromatic hydrocarbons from aromatic/aliphatic mixtures. *Fuel Process. Technol.,* **2005**, *87*(1), 59-70.
        [http://dx.doi.org/10.1016/j.fuproc.2005.06.002]

[25]   Meindersma, G.W.; Podt, A.J.G.; Klaren, M.B.; de Haan, A.B. (J. G.); Klaren, M. B.; de Haan, A. B.
        Separation of aromatic and aliphatic hydrocarbons with ionic liquids. *Chem. Eng. Commun.,* **2006**,
        193(11), 1384-1396.
        [http://dx.doi.org/10.1080/00986440500511403]

[26]   Meindersma, G.W.; de Haan, A.B. Conceptual process design for aromatic/aliphatic separation with
        ionic liquids. *Chem. Eng. Res. Des.,* **2008**, *86*(7), 745-752.

[http://dx.doi.org/10.1016/j.cherd.2008.02.016]

[27]   de Riva, J.; Ferro, V.R.; Moreno, D.; Diaz, I.; Palomar, J. Aspen Plus Supported Conceptual Design of the Aromatic–Aliphatic Separation from Low Aromatic Content Naphtha Using 4-Methyl-N-Butylpyridinium Tetrafluoroborate Ionic Liquid. *Fuel Process. Technol.,* **2016**, *146*, 29-38.
[http://dx.doi.org/10.1016/j.fuproc.2016.02.001]

[28]   Zhang, J.; Huang, C.; Chen, B.; Ren, P.; Lei, Z. Extraction of Aromatic Hydrocarbons from Aromatic/Aliphatic Mixtures Using Chloroaluminate Room-Temperature Ionic Liquids as Extractants. *Energy Fuels,* **2007**, *21*(3), 1724-1730.
[http://dx.doi.org/10.1021/ef060604+]

[29]   Pereiro, A.B.; Rodríguez, A. An Ionic Liquid Proposed as Solvent in Aromatic Hydrocarbon Separation by Liquid Extraction. *AIChE J.,* **2010**, *56*, 381-386.
[http://dx.doi.org/10.1002/aic.11937]

[30]   Navarro, P.; Larriba, M.; García, J.; Rodríguez, F. Design of the recovery section of the extracted aromatics in the separation of BTEX from naphtha feed to ethylene crackers using [4empy][Tf$_2$N] and [emim][DCA] mixed ionic liquids as solvent. *Separ. Purif. Tech.,* **2017**, *180*, 149-156.
[http://dx.doi.org/10.1016/j.seppur.2017.02.052]

[31]   Shaahmadi, F.; Hashemi Shahraki, B.; Farhadi, A. Liquid–liquid extraction of toluene from its mixtures with aliphatic hydrocarbons using an ionic liquid as the solvent. *Sep. Sci. Technol.,* **2018**, *53*(15), 2409-2417.
[http://dx.doi.org/10.1080/01496395.2018.1449859]

[32]   Salleh, M.Z.M.; Hadj-Kali, M.K.; Hashim, M.A.; Mulyono, S. Ionic liquids for the separation of benzene and cyclohexane – COSMO-RS screening and experimental validation. *J. Mol. Liq.,* **2018**, *266*, 51-61.
[http://dx.doi.org/10.1016/j.molliq.2018.06.034]

[33]   Yao, C.; Hou, Y.; Sun, Y.; Wu, W.; Ren, S.; Liu, H. Extraction of aromatics from aliphatics using a hydrophobic dicationic ionic liquid adjusted with small-content water. *Separ. Purif. Tech.,* **2020**, *236*, 116287.
[http://dx.doi.org/10.1016/j.seppur.2019.116287]

[34]   Du, W.; Slaný, M.; Wang, X.; Chen, G.; Zhang, J. The Inhibition Property and Mechanism of a Novel Low Molecular Weight Zwitterionic Copolymer for Improving Wellbore Stability. *Polymers (Basel),* **2020**, *12*(3), 708.
[http://dx.doi.org/10.3390/polym12030708] [PMID: 32210118]

[35]   Du, W.; Wang, X.; Chen, G.; Zhang, J.; Slaný, M. Synthesis, Property and Mechanism Analysis of a Novel Polyhydroxy Organic Amine Shale Hydration Inhibitor. *Minerals (Basel),* **2020**, *10*(2), 128.
[http://dx.doi.org/10.3390/min10020128]

[36]   Rahman, M.T.; Negash, B.M.; Moniruzzaman, M.; Quainoo, A.K.; Bavoh, C.B.; Padmanabhan, E. An Overview on the potential application of ionic liquids in shale stabilization processes. *J. Nat. Gas Sci. Eng.,* **2020**, *81*, 103480.
[http://dx.doi.org/10.1016/j.jngse.2020.103480]

[37]   An, Y.; Jiang, G.; Ren, Y.; Zhang, L.; Qi, Y.; Ge, Q. An environmental friendly and biodegradable shale inhibitor based on chitosan quaternary ammonium salt. *J. Petrol. Sci. Eng.,* **2015**, *135*, 253-260.
[http://dx.doi.org/10.1016/j.petrol.2015.09.005]

[38]   Yang, L.; Jiang, G.; Shi, Y.; Yang, X. Application of Ionic Liquid and Polymeric Ionic Liquid as Shale Hydration Inhibitors. *Energy Fuels,* **2017**, *31*(4), 4308-4317.
[http://dx.doi.org/10.1021/acs.energyfuels.7b00272]

[39]   Yang, L.; Yang, X.; Wang, T.; Jiang, G.; Luckham, P.F.; Li, X.; Shi, H.; Luo, J. Effect of Alkyl Chain Length on Shale Hydration Inhibitive Performance of Vinylimidazolium-Based Ionic Liquids. *Ind. Eng. Chem. Res.,* **2019**, *58*(20), 8565-8577.
[http://dx.doi.org/10.1021/acs.iecr.9b01016]

[40]   Li, X.; Jiang, G.; Yang, L.; Wang, K.; Shi, H.; Li, G.; Wu, X. Application of Gelatin Quaternary Ammonium Salt as an Environmentally Friendly Shale Inhibitor for Water-Based Drilling Fluids. *Energy Fuels,* **2019**, *33*(9), 9342-9350.
       [http://dx.doi.org/10.1021/acs.energyfuels.9b01798]

[41]   Jia, H.; Huang, P.; Wang, Q.; Han, Y.; Wang, S.; Dai, J.; Song, J.; Zhang, F.; Yan, H.; Lv, K. Study of a gemini surface active ionic liquid 1,2-bis(3-hexylimidazolium-1-yl) ethane bromide as a high performance shale inhibitor and inhibition mechanism. *J. Mol. Liq.,* **2020**, *301*, 112401.
       [http://dx.doi.org/10.1016/j.molliq.2019.112401]

[42]   Joshi, V.A.; Kundu, D. Ionic liquid promoted extraction of bitumen from oil sand: A review. *J. Petrol. Sci. Eng.,* **2021**, *199*, 108232.
       [http://dx.doi.org/10.1016/j.petrol.2020.108232]

[43]   Painter, P.; Williams, P.; Mannebach, E. Recovery of Bitumen from Oil or Tar Sands Using Ionic Liquids. *Energy Fuels,* **2010**, *24*(2), 1094-1098.
       [http://dx.doi.org/10.1021/ef9009586]

[44]   Li, X.; Sun, W.; Wu, G.; He, L.; Li, H.; Sui, H. Ionic Liquid Enhanced Solvent Extraction for Bitumen Recovery from Oil Sands. *Energy Fuels,* **2011**, *25*(11), 5224-5231.
       [http://dx.doi.org/10.1021/ef2010942]

[45]   Tian, Y.; McGill, W.B.; Whitcombe, T.W.; Li, J. Ionic Liquid-Enhanced Solvent Extraction for Oil Recovery from Oily Sludge. *Energy Fuels,* **2019**, *33*(4), 3429-3438.
       [http://dx.doi.org/10.1021/acs.energyfuels.9b00224]

[46]   Tourvieille, J.N.; Larachi, F.; Duchesne, C.; Chen, J. NIR hyperspectral investigation of extraction kinetics of ionic-liquid assisted bitumen extraction. *Chem. Eng. J.,* **2017**, *308*, 1185-1199.
       [http://dx.doi.org/10.1016/j.cej.2016.10.010]

[47]   Pulati, N.; Lupinsky, A.; Miller, B.; Painter, P. Extraction of Bitumen from Oil Sands Using Deep Eutectic Ionic Liquid Analogues. *Energy Fuels,* **2015**, *29*(8), 4927-4935.
       [http://dx.doi.org/10.1021/acs.energyfuels.5b01174]

[48]   Zhao, Y.; Paso, K.; Zhang, X.; Sjöblom, J. Utilizing ionic liquids as additives for oil property modulation. *RSC Advances,* **2014**, *4*(13), 6463.
       [http://dx.doi.org/10.1039/c3ra46842a]

[49]   Zheng, C.; Brunner, M.; Li, H.; Zhang, D.; Atkin, R. Dissolution and suspension of asphaltenes with ionic liquids. *Fuel,* **2019**, *238*, 129-138.
       [http://dx.doi.org/10.1016/j.fuel.2018.10.070]

[50]   Bera, A.; Agarwal, J.; Shah, M.; Shah, S.; Vij, R.K. Recent advances in ionic liquids as alternative to surfactants/chemicals for application in upstream oil industry. *J. Ind. Eng. Chem.,* **2020**, *82*, 17-30.
       [http://dx.doi.org/10.1016/j.jiec.2019.10.033]

[51]   EL-Hefnawy, M.E.; Atta, A.M.; El-Newehy, M.; Ismail, A.I. Synthesis and characterization of imidazolium asphaltenes poly (ionic liquid) and application in asphaltene aggregation inhibition of heavy crude oil. *J. Mater. Res. Technol.,* **2020**, *9*(6), 14682-14694.
       [http://dx.doi.org/10.1016/j.jmrt.2020.10.038]

FINAL CONCLUSIONS AND FUTURE PROSPECTS

Many studies on the application of ILs in the Oil Industry have been described in this book. Due to the unique properties of ILs, this class of materials has been considered as a very attractive alternative to be used in many fields and the obtained research results are very attractive in most of the applications.

On the other hand, the possibility of synthesizing a large number of ILs with physicochemical properties such as acid-base, viscosity, and solubility that can be modulated makes this family very interesting to attend different problematics and processes.

In spite of the fact that the effectiveness of ILs has been proven in the laboratory through various types of experiments with excellent results, there still exists a gap in demonstrating the application of ILs on high scale. From those described in this book, the use of ILs to produce alkylation gasoline is the only application that has been demonstrated on a large scale with commercial technologies available at high level such as Ionylation and Isoalky. Even in these cases, the technologies are still limited in their application by catalyst chemical stability problems (high moisture sensitivity) that characterize ILs with Lewis acidic properties.

One of the advantages of ILs is their practically zero vapor pressure, which apparently makes them environmentally friendly solvents and ideal substituents of volatile organic solvents in the development of sustainable processes; however, this issue must be faced rigorously, since many ILs have high toxicity and low biodegradability problems.

Another problem that plagues most ILs is their high cost. Currently, there are already suppliers of ILs in large volumes and this limitation will be solved as more efficient and sustainable ILs are developed.

Notwithstanding, in our opinion, the limitations that currently exist regarding the industrial-scale applications of ILs may be resolved as research progresses and is carried out through a more comprehensive approach. Many research works are under progress to reduce the production cost of ILs and develop processes based on ILs synthesized from natural and biodegradable raw materials; likewise, truly efficient, economical, and sustainable regeneration processes are being conceived. More refined screening methods will have to be developed for the selection of an IL for a specific application. With all these considerations, we hope that in the near future, more and more large-scale uses of ILs in the chemical and oil industries will be part of the aforementioned success stories.

SUBJECT INDEX

A

Acid(s) 12, 14, 15, 21, 24, 25, 27, 28, 30, 43, 44, 48, 83, 61, 77, 97, 99, 101, 109, 110, 114, 115, 149, 160, 162, 165, 166, 169, 170, 172, 173
  acetic 21
  amino (AAs) 43, 44, 48, 83, 97
  anthranilic 48
  formic 21
  hydrochloric 110
  hydrofluoric 27, 28
  inorganic 172
  lewis 30, 160, 166, 170, 173
  naphthenic 30, 77
  sulfhydric 149
  sulfuric 28, 109, 110, 114, 115, 165
  rain 12, 14, 15
  trifluoromethanesulfonic 165
Activity 3, 112, 115, 169
  anticorrosive 115
  hydrophilic 112
Additional oil recovery tests 142
Adsorption 29, 39, 81, 103, 109, 110, 111, 114
  effect 114
  process 110, 111
Agents 23, 60, 63, 67, 68, 69, 88, 101, 125, 128, 133, 142, 146
  anti-agglomerant 125, 128, 133
  demulsifying 60, 63, 67, 68, 69, 88
  multifunctional 146
  oxidizing 23, 101
Alkali-surfactant-polymer (ASP) 141, 149
Alkylation 6, 27, 28, 29, 160, 161, 162, 163, 164, 165, 166, 169, 170, 172, 173
  gasoline 27, 28, 29, 160, 161, 163, 172, 173
  process 6, 27, 29, 161, 162, 164, 170
  reaction 6, 160, 162, 163, 164, 165, 166, 169, 170, 172, 173
Amino 43, 85, 186
  polyether 186

synthetic 85
Anions 2, 3, 19, 20, 42, 43, 44, 48, 50, 67, 68, 86, 113, 114, 115, 166, 168, 173, 182, 190
  anthranilate 50
  halo aluminate 173
  halogenated 166
  long-chain carboxylate 182
  organic acid-based 190
  polyatomic 86
Anti-agglomerant effect 128
Anti-caking effect 128
Anticorrosion program 101
Applications, industrial-scale 45
Asphaltenes-ionic liquids interactions 87
Atmospheric pollution 14
Atomic force microscopy (AFM) 114
Attraction forces 60, 114
  electrostatic 114
Automotive vehicles 14

B

Behavior 106, 129, 131, 134
  kinetic inhibition 131
Biosurfactants, cellobioside 83
Brownian motion forces 62

C

Carboxylic 77, 125
  acids 77
  anhydrides 125
Catalysts, synergistic 164
Cations 2, 108, 110, 111, 132, 133, 141, 150
  heterocyclic 111
  ionic monomer 150
  organic 2, 141
  pyridinium 108, 132, 133
  quaternary ammonium 110

X

www.ingramcontent.com/pod-product-compliance
Lightning Source LLC
Chambersburg PA
CBHW050846220326

41598CB00006B/447